MORE THAN NATURE NEEDS

Derek Bickerton

More than Nature Needs

Language, Mind, and Evolution

HARVARD UNIVERSITY PRESS
Cambridge, Massachusetts, and London, England   2014

Copyright © 2014 by the President and Fellows of Harvard College
All rights reserved
Printed in the United States of America

Library of Congress Cataloging-in-Publication Data

Bickerton, Derek.
More than nature needs : language, mind, and evolution / Derek Bickerton.
   pages   cm
Includes bibliographical references and index.
ISBN-13: 978-0-674-72490-7
 1. Language and languages.   2. Human evolution—Psychological aspects.
3. Language acquisition—Psychological aspects.   4. Cognitive grammar.
5. Psycholinguistics.   I. Title.
P106.B468 2014
 401'.9—dc23   2013011146

For Yvonne
  To hold in age's calm
  All the sweets of youth's longings . . .

Contents

1. Wallace's Problem   *1*
2. Generative Theory   *16*
3. The "Specialness" of Humans   *42*
4. From Animal Communication to Protolanguage   *73*
5. Universal Grammar   *109*
6. Variation and Change   *151*
7. Language "Acquisition"   *185*
8. Creolization   *218*
9. *Homo Sapiens Loquens*   *258*

References   *279*
Acknowledgments   *317*
Index   *319*

O reason not the need! Our basest beggars
Are in the poorest thing superfluous.
Allow not nature more than nature needs,
Man's life is cheap as beast's.
—Shakespeare, *King Lear,* Act 2, Scene 4

It's a possibility that there's something we just don't fundamentally understand, that it's so different from what we're thinking about that we're not thinking about it yet.
—Leonard Krugelyak, *Nature* 456.21 (2008)

CHAPTER 1

Wallace's Problem

The structure of this book is simple. In this chapter I state a problem and outline what I think is its solution. The rest of the book consists of arguments and evidence that support this solution. The problem itself, though quite easy to state, has ramifications that will take us through the territories of a number of disciplines, including evolutionary biology, paleoanthropology, psychology, neurobiology, and linguistics. Rest assured that everything will eventually lead us back to this same question, one of the most crucial anyone can ask: How did the human species acquire a mind that seems far more powerful than anything humans could have needed to survive?

Since it is becoming a custom to name problems after people (Plato's problem, Darwin's problem, Orwell's problem, etc.), let us call this problem Wallace's problem, since it was Alfred Russel Wallace, cofounder with Darwin of the theory of evolution through natural selection, who was the first to state it clearly and unequivocally. In his own words, "Natural selection could only have endowed the savage with a brain a little superior to that of an ape whereas he possesses one very little inferior to that of an average member of our learned societies" (Wallace 1869: 204). By "savage," the customary expression of the time, Wallace meant only someone who had had what many nowadays would consider the good fortune to be born into a preliterate, pre-industrial society. His estimate of "savage" intellectual capacity was actually pretty enlightened for that time—decades would pass before anyone had the honesty to replace "very little inferior" with "equal." And yet recognizing the universality of human intelligence gave Wallace only disquiet.

2 Wallace's Problem

If evolution was a gradual process, and natural selection responded only to the demands placed on animals by their environment, then humans *should* have had a brain "little superior to that of an ape." A brain slightly better than an ape's would have enabled them to outsmart anything else on two legs or four, to reach the top of the food chain. Early humans didn't need to do math, build boats, compose music, or have ideas about the nature of the universe in order to do all the things early humans did. That they should suddenly find themselves endowed with brains that could potentially enable them to do all these things was remarkable enough. But more remarkable yet was the fact that those same brains would make it possible for their possessors to cover the entire world with their works, to plunge into the deepest depths of the ocean, to soar into the highest reaches of the atmosphere, and (less that half a century after Wallace's death) to leave even the Earth itself behind.

Wallace couldn't bring himself to believe that natural selection alone could have done all this. There must have been some form of supernatural influence involved in the sudden and abrupt creation of the immense gap between human mental abilities and those of any other species. This gap seemed especially remarkable because nothing similar existed anywhere else in nature. What appeared elsewhere was exactly what any theory of evolution through natural selection would have predicted: isolated islands of highly task-specific adaptation, backed by otherwise smooth gradations of cognitive capacity across the entire range of species, leaving only humans as remote and exotic outliers.

Many writers on the history of evolutionary theory have attributed Wallace's views on human evolution to his conversion to spiritualism—a good way of making his problem disappear. But regardless of what Wallace believed, his problem remains. The human mind *is* a profoundly unlikely evolutionary development, from any perspective, and we should honor Wallace's honesty in facing this problem, regardless of how we feel about the solution he proposed.

Though Wallace was the first to clearly articulate the problem, it is almost certainly something that earlier minds were in some sense aware of. When Shakespeare wrote the lines that form the first epigraph to this book, he was purportedly expressing no more than Lear's anger at his daughter for limiting the number of his attendants. But with Shakespeare there is always layer upon layer of meaning beneath the lines (one reason he is the greatest of writers). Underlying these particular lines is his awareness that even the "basest" of humans had far more than was needed for purely material purposes and that the lives of "beasts" are, in comparison with ours, far more limited. How this could have come about without the intervention of any mysterious extra-evolutionary forces is the topic of this book.

Darwin's Response

Darwin certainly realized what the problem was. He had "no doubt that the difference between the mind of the lowest man and that of the highest animal is immense" (Darwin 1871: 100). He repeated Wallace's estimate of "savages," pointing out that the three natives of Tierra del Fuego who had accompanied him on the *Beagle* "resembled us in disposition, and in most of our mental faculties." However, at the same time he ingeniously disarmed the argument from the gap between ape and human by citing against it the continuous gradation of intellect across the "much wider interval in mental powers" between "the lower fishes" and "the higher apes." If there was a gradation in the one case, then there must, contrary to appearances, be a gradation in the other, since "there is no fundamental difference of this kind" (34).

This is sheer sleight of hand. The gradation of intellect between lamprey and chimpanzee is an argument not *against* the gap but *for* it. If there are countless species with abilities partway between those of lamprey and chimp, there should also be many species intermediate between chimps and humans. How is it that there are no animals with small or moderate amounts of self-consciousness, gradually increasing degrees of innovation and creativity, varying levels of artistic achievement (perhaps in only one or two of the arts), or at least a rudimentary language? The flat assertion of "no fundamental difference" is not (and could not have been, even in Darwin's time) a scientific statement. It was and is a pure declaration of faith.

Darwin sought to give empirical backing to this declaration by the same means he used to support his claims in *The Origin of Species:* by accumulating a large stock of mostly anecdotal reports of behavior in other animals. But what is valid where there is also objective evidence in the physical forms of the various species involved is much less so when mental capacities are at issue. Since there is no unambiguous objective evidence to support these anecdotes, subjective interpretations, notoriously variable and unreliable, have to be unquestioningly relied on. A widespread human tendency to anthropomorphize puts its stamp on far too much of the evidence here.

Yet even here Darwin, ever a cautious thinker, hedges his bet. He continues to profess his faith that "the difference in mind between man and the higher animals, great as it is, is certainly one of degree and not of kind." But the examples he cites in the same paragraph all involve emotions rather than cognitive processes. He feels forced to immediately suggest a fall-back position: "If it be maintained that certain powers, such as abstraction, self-consciousness etc., are peculiar to man, it may well be that these are the incidental results of other highly-advanced intellectual faculties; and these again are mainly the result of the continued use of a highly-developed language" (1871: 103).

This was a brilliant insight, but in Darwin's day it could not have been more than a promissory note. Darwin did not and in the second half of the nineteenth century could not have cashed it out even in terms of hypothetical proposals, let alone plausible mechanisms backed by empirical evidence. Besides, to him it was simply Plan B, something he confidently believed he would never need. Future research, he must have thought, would surely spell out in detail the missing pieces, the discoveries that would show animal powers to be really only a little less than human.

And that, for a century, was that. No one attempted to solve Wallace's problem. No one would even admit that there was such a thing, which saved them from the difficult if not impossible task of explaining why it *wasn't* a problem. Of course if you were a creationist or believed in any form of spiritual intervention, there was no problem. God, or the Life Force, just did it. Nothing illustrates the intellectual incapacity of creationists and believers in intelligent design better than their almost total failure to exploit this issue. Repeatedly in their literature these groups claim as one of their strongest arguments against evolution the "fact" that no form intermediate between apes and humans has been found. Only a mouse-click away from them are a score of sites where they would learn that, far from an absence, there is an embarrassing number of intermediate forms, providing endless fuel for paleontological argument—not about whether these are really intermediates but simply over issues like whether they are directly ancestral to humans or on a side branch, whether specimen X should be assigned to species A or B, or whether, for that matter, A and B shouldn't be merged (except for those who maintain that B should be subdivided into species C and D). In other words, normal science in progress. But creationists and designers alike have focused almost exclusively on physical form, where there is abundant evidence for evolutionary continuity, rather than on cognitive behavior, where there is little or none.

Evolutionists should be properly grateful for this misdirection. The cognitive gap between humans and nonhumans is evolution's Achilles' heel. Wallace's problem is real, and evolutionists have simply ignored it or tried to explain it away. The only author I know of who has restated it is David Premack (1986: 133), who noted that "human language is an embarrassment for evolutionary theory because it is vastly more powerful than one can account for in terms of selective forces." Everyone else has simply repeated, in one form or another, the mantra that humans are just "another unique species" (Foley 1987). Researchers have assembled massive lists—this time based on much more than anecdotal evidence—of all the clever things that other animals can do (see, e.g., some of the commentaries on Penn et al. 2008).

In some quarters it has become politically incorrect even to mention all the clever things that humans can do and animals can't. But there can be no question that such things exist. "Human animals—and no other—build fires and wheels, diagnose each other's illnesses, communicate using symbols, navigate with maps, risk their lives for ideals, collaborate with each other, explain the world in terms of hypothetical causes, punish strangers for breaking rules, imagine impossible scenarios, and teach each other how to do all of the above" (Penn et al. 2008: 109). If it is true that "for over 35 years, researchers have been demonstrating through tests both in the field and in the laboratory that the capacities of nonhuman animals to solve complex problems form a continuum with those of humans" (Pepperberg 2005: 469), how can this be? One's initial reaction may well be that both statements can't be true. But they are. This is just one of the paradoxes that Wallace's problem forces us, or should force us, to face.

6   **Wallace's Problem**

In practice it doesn't, because after Premack rudely resurrected the problem, silence followed. Even Noam Chomsky, who for many years had insisted that language at least was totally divorced from anything other animals did, finally entered the fold and accepted the conventional wisdom that other animals have, among them or between them, all the bits and pieces required for language except perhaps one (Hauser et al. 2002; Chomsky 2007). But surely it can't be long before some creationist or believer in intelligent design catches on to the fact that Wallace's problem is the ideal place for inserting the "wedge" that creationists are always talking about (Johnson 1997). If that happens, science will find itself in a very embarrassing position, because as of now a scientific solution to Wallace's problem just doesn't exist.

The Key to the Problem

The key to the problem lies in Darwin's Plan B, cited above: "It may well be that [abstraction, self-consciousness, etc.] are the incidental results of other highly-advanced intellectual faculties; and these again are mainly the result of the continued use of a highly-developed language."

This notion, if seriously advanced, is bound to meet with consumer resistance. For many people, language is merely "a means of communication." Like the Morse code or semaphore flags, it is not in itself constitutive of meaning but merely transmits meanings that have come from somewhere else. First you must "have a thought," which you then dress up in words, though exactly what a thought is and where it comes from is far less clear than what words are and where they come from. For such persons, Darwin had it backward. First human intelligence must have developed, and only after long development could anything like language have emerged.

"Intelligence" has proved extremely difficult to define, and we would have a hard task on our hands if we were asked to find another species whose particular abilities derived from any kind of superior "general intelligence" that set them apart from other species. In most cases it's the other way around. In other species, it's glaringly obvious that cognitive skills are geared to very specific behaviors: spider intelligence to web spinning, beaver intelligence to dam building, bat intelligence to echolocation, bee intelligence to pollen gathering, and so forth. In the case of apes, there might seem to be a less narrowly focused intelligence than in the species I've mentioned. It's as if social intelligence, notoriously their strong suit, had spilled over into other areas—a natural enough development, since social intelligence has to be more flexible, able to cope with constantly changing roles and status levels, if it is to work at all. It is also true that apes, like dogs but even more so, can be trained by humans to do a few of the things that otherwise only humans do. But apes, left to themselves, have never innovated the kinds of thing they are taught, whereas humans, unless taught by space aliens, must have spontaneously produced innovations over and over again.

What all this means is that, in terms of evolution, any "increase in intelligence" that is not motivated by the specific demands of a particular species' niche is highly unlikely, perhaps even an impossibility. What most often, perhaps always drives increase of intelligence is the development of some very specific ability that is required if the species is to solve an ecological problem, such as how to catch fast-moving and skillfully maneuvering flying insects during the hours of night. The way increases of intelligence arise forms only one aspect of an even more general evolutionary process, covering every aspect of form and behavior: "The diversity of species . . . represents variation in design suggestive of *adaptation to specific tasks*" (Weibel 1998: 1, emphasis added).

Normally, specialized intelligence doesn't spread to inform other areas precisely because it has to be focused on a very narrow range of behaviors. (Gardner [1983], with his "multiple intelligences," and evolutionary psychology's "Swiss army knife" approach to intelligence [e.g., Barkow et al. 1992] deal with similar issues from somewhat different perspectives.) But in very rare cases an initially focused kind of intelligence may be able to spread.

What is still lacking is any understanding of how and why and under what set of circumstances a focused intelligence could spread, and especially how all this could have come about in the specific case of humans and only among humans. The kind of answer one gets is all too often along the lines of this: "One *possibility* . . . is that recursion in animals represents a modular system designed for a particular function (e.g. navigation) and impenetrable with respect to other systems. During evolution, [this system] *may have* become penetrable and domain-general. . . . This change from domain-specific to domain-general *may have* been guided by particular selective pressures, unique to our evolutionary past, or as a consequence (by-product) of *other kinds* of neural re-organization" (Hauser et al. 2002: 1578, emphasis added). Such pronouncements merely restate the problem in a more complicated way without shedding any light on it.

We will have to do a lot better than that. Having identified the key source of human intelligence, we will have to show how that particular source was able to create the thoroughly convincing illusion that human "higher powers" spring from possession of some overarching, all-purpose intelligence.

But if, following Darwin, that key source is identified as language, we face serious obstacles. Language is a clear candidate, of course, because it is a very specific and specialized form of behavior and because the mechanisms through which it operates are clearly identifiable and well-studied (especially as compared with things like abstraction, consciousness, foresight, or imagination). Thus it fits the normal evolutionary profile of how increases in intelligence come about. However, it seems to yield no clues as to how that intelligence could have generalized across the entire spectrum of behaviors. Thus either Darwin's Plan B is wrong, or we do not yet have any adequate understanding of what language is and does. I would opt for the second of these choices.

This claim may seem both arrogant and misguided, given the existence of a whole slew of linguistic theories worked out by dedicated professionals. It would indeed be so but for one fact: theories of language, without exception, have been worked out on the basis of synchronic linguistic evidence, without paying any attention to language as an outcome of evolutionary processes. All theories of language fall into one of only two classes. One class holds that there are few if any biological adaptations (and those mostly confined to speech) that are devoted exclusively to language, which consequently had to arise as a result of more general cognitive developments. The other class holds that, apart from peripheral elements, language (and in particular, syntax) is determined by highly task-specific (if yet to be specified) biological mechanisms.

In terms of evolutionary biology, both approaches are equally implausible. On the one hand, any theory that denies or minimizes the role of a task-specific genetic infrastructure for language ignores the fact that no other significant species-specific trait in the whole of nature lacks such infrastructure. On the other hand, any theory that exaggerates the size and/or specificity of linguistic infrastructure ignores the fact that for no other significant species-specific trait does the genetic infrastructure spell out not just the basic processes required by that trait but all possible variations in those processes. Yet that is exactly what is required by a Chomskyan Language Acquisition Device (LAD; Chomsky 1965). Such a device must be even-handed. It cannot favor one kind of language over another; it must make equally possible the learning of each of earth's several thousand languages.

Instead of these alternatives, any biologist who came to the study of language with no philosophic or linguistic preconceptions would predict a genetic component that provided the minimum basic mechanisms necessary for language and left subsequent variation to environmental factors. Only an approach along these lines can explain both why humans, but no other species, have language and why languages, while following the same *Bauplan,* vary unpredictably from one another in structural details. Only such an approach can integrate the evolution of language into an overall account of human evolution. And only by taking such an approach—one that no current theory of language has taken—can we show how nature could have provided our species with powers so far in excess of their needs.

But to fully understand the common birth of language and human cognition we have first to get language evolution right. If we don't know how language evolved we don't really know what it is, or what its real properties are, or why it should have those properties and not others.

Finding out how language evolved has been called "the hardest problem in science" (Christiansen and Kirby, 2003). This is an extreme statement but does not seem an unreasonable one, given that attempted solutions go back for centuries and comprise countless theories, hypotheses, and sheer conjectures. If the statement is incorrect, as it is, that is not the fault of Christiansen and Kirby. For why language evolution *isn't* the hardest problem in science is not because it isn't hard. It's because it isn't a problem. It's three problems.

This is the single most important point to grasp about the whole issue of language evolution. It is also the point in which this book differs most sharply from all previous work in the field, as well as being what made the problem look so hard in the first place. Once we see that its difficulty came from mixing apples with oranges, compounding three totally different processes under the rubric of "language evolution," the picture changes radically. Now instead of the "hardest problem"—"the insoluble problem," we should have called it—we now find ourselves faced with three problems, each of which, once it has been distinguished from the others, is relatively easy to solve. What is more, adopting a trio of different solutions not only makes for an explanation of language evolution that aligns it more closely with evolutionary developments in other species; it also takes us beyond the sterile and seemingly unending arguments of empiricists and nativists alike.

The very persistence of those arguments should have told us something. Scientific debates seldom last so long unless the issues have not yet been correctly formulated, so that the wrong questions are being asked. Language itself, too, seemed to be telling us that no single viewpoint would suffice to give a full account of it. Its Janus-faced nature was succinctly expressed by Deacon (2010: 9005): "Language is too complex and systematic, and our capacity to acquire it is too facile, to be adequately explained by cultural use and general learning alone. But the process of evolution is too convoluted and adventitious to have produced this complex phenomenon by lucky mutation or the genetic internalization of language behavior." Unfortunately, he too persisted with a single factor—relaxed selection—in trying to explain how language evolved.

Instead, the process must be broken into three processes, each of which requires separate questions and separate answers. The first process is escape from the prison of animal communication, and the question is how and why this escape was accomplished and what were the factors involved. The second process is the acquisition of very basic structures for dealing with the output of the first process, and again we must ask how and why and with what agency this structure-building was executed. The third process is the creation of the kind of language we know today; once more the question is one of how and why when true language emerged it took the form it did, and what (or who) made it take that form

Proposing answers to these questions enables us to see why language should have the dual aspect characterized by Deacon. We can also see why the nativist-empiricist debate has persisted for so long. Recall the blind men of Hindustan who, having touched different parts of an elephant, decided it was a variety of things—a snake, a spear, a tree-trunk and so on. In just the same way, most language evolutionists had their hands on different parts of the elephant, and drew from it wrong conclusions about the nature of the beast, mistaking part for whole.

All had grasped a part of the truth. Natural selection, internal development, and culture have all played roles in the evolution of language. It's just that they haven't played them at the same time or in the same process. In the first process, natural selection played the most prominent role. In the second, it was internal development. In the third, it was culture.

The Organization of This Book

To provide the reader with a compass for the rest of the book, I briefly outline what I think are the reasons for the three processes and the means by which they were realized. The first was driven mainly by external evolutionary factors. Like web-spinning, echolocation, dam-building and all the other behavioral innovations we know of, it arose not through unmotivated genetic changes but as a direct result of a specific ecological problem that the species concerned—one ancestral to humans, around two million years ago—had to face. As often happens, response to an ecological problem involved constructing a new ecological niche (Odling-Smee et al. 2003), one that mandated referential displacement (the capacity to transfer information about entities and events that lay outside the immediate sensory range of the animals concerned) in order to develop and fully exploit the niche. What resulted from displacement was no more than an enhanced form of animal communication. Human ancestors had acquired brains only a little better than those of apes; Wallace would have been happy with that.

In ants or bees, that would have been the whole story. In a species with brains orders of magnitude larger than those of ants and bees, the story couldn't stop there. In such a species, the continued use of displaced reference had to extend to areas outside the immediate foraging function of the enhanced communication. The more they were used, for the more purposes, the more these displaced units—proto-words, we might as well call them—would approximate more closely to fully

12   Wallace's Problem

symbolic units, and the neural representation of each unit would be linked with a (presumably pre-existing) concept. This opened the path for a quite separate series of developments, one that Wallace did not know and could not have known about, because the necessary understanding of how brains operate still lay a hundred years in the future.

The presence in the brain of representations of symbolic units set the second process in motion. Brains have been self-organizing and self-reorganizing for hundreds of millions of years, in order to deal with information streams coming from the outside world through the various senses. However, in the case of human ancestors the brain also had to respond to a growing store of words and their associated concepts, and the information streams that these phenomena created. To respond, the brain had to re-organize its resources, just as brains had had to do every time an animal developed a new sense. Brain restructuring is driven not by selective pressures from the environment but from the brain's own need to economize energy and function with a minimal amount of wiring. It had to allot or re-allocate spaces for storage of proto-words. It had to redraw its wiring diagram so as to link words with their appropriate concepts and with one another and with the motor controls for speech. Since its owners made blundering efforts to string both proto-words and concepts together for its utterances and thoughts respectively, it was obliged to develop simple stereotypical and fully automated routines to reduce the effort of assembling them each time on the fly. It had done no less for other repeated behaviors: throwing, picking up, striking, etc.

The syntactic engine that was thus developed knew nothing of thoughts or sentences. That distinction, vital to humans, is immaterial to the brain. The brain knows only cell-complexes that store information and neural impulses that link the information and can create larger units from it. Those units may not travel outside the brain; they are then thoughts. Or they may be externalized via the motor organs of speech, after another round of grammatical processing, in which case they become sentences. Originally there was no further round of grammatical processing. How one came to be is the story of the third process.

The automated algorithms that assembled both words and thoughts were perfectly adequate for thinking with. In both thought and language you need to know the precise relations between the things you're talking or thinking about—who (or what) did what to who (or what), for whom (or what), with what, how, when, where and very often why. If you don't get those relations right, all you've got is word-salad or thought-salad. But if what you're doing is thinking, you yourself already *know* those relations; they don't have to be spelled out. And if you're talking, the same applies. But if you're talking, someone is usually listening, and that someone *doesn't* know and can't know unless such relations are overtly spelled out. That's a problem for humans. It's not in any way a problem for brains.

Imagine you are a human brain. What would you care whether the lumbering life-support system that carries you around can easily decode the ingenious product you had made for it? That life-support system had made life hard for you by inventing thousands of words you were expected to store for it and by insisting on stringing those words together and pushing them out of its mouth. You had loyally stored the words so that they were instantly accessible and created algorithms that automatically gave shape to the word/thought salads that were all the clumsy brute could manage on its own. Now it says it can't process the stuff? Enough already!

So the brain played no active role in the third process. Humans, left to their own devices, initiated it. We need not suppose that there was anything conscious or intentional about this. Doubtless at first they got along as best they could with the (very) bare original syntax. But everything varies and consequently changes, and every now and then some accidental feature would be found to help in disambiguating structure. Such features might then take on a life of their own, as it were, sometimes with a runaway effect leading to such phenomena as Finnish or Hungarian cases, Chinese noun-classifiers, or (in some Bantu languages) a single morpheme you have to attach to every word in a clause. Note that the behavior of what one might call low-level syntactic phenomena is strongly lineage-dependent (Dunn et al. 2011); a fact inexplicable by any theory of universals, whether Chomskyan or Greenbergian, though it follows directly from what is proposed here.

As noted above, viewing the evolution of language as three quite separate processes renders moot the sterile debates over empiricist versus nativist theories and internalist versus externalist explanations. But it also superannuates equally sterile debates over the relations between language and thought, such as whether a "language of thought" exists or is even possible. Previously, the latter debate was vitiated by the notion that "thinking in language" could only mean thinking in one of the several thousand actual languages that are known. Once it is realized that "thinking in language" means thinking with the bare output of the syntactic engine, something lacking in all the features that make one language different from another, the debate simply evaporates.

Not only does it evaporate, but we can also appreciate the reason for a seldom noticed but indisputable fact. On the surface at least, languages differ widely. Nobody speaks alike, but apart from some minor differences in a few quite limited semantic areas, everybody thinks alike. Language in America is not at all like language in Zimbabwe, and neither one is like language in China. But Americans, Chinese, and Zimbabweans all use identical rational processes and (if we subtract cultural differences, which nowadays are rapidly subtracting themselves, anyway) think similar thoughts. This is a natural result, given a theory that regards syntacticized language as neither fully innate nor fully learned, but rather compounded of a learned component and an innate component, with only the latter used in thinking.

However, we must keep priorities in mind. The ultimate goal of this book is to solve Wallace's problem, to show how in a very few million years a not particularly distinguished primate could have so far exceeded the capacities of all other animals through normal evolutionary processes. But this goal is predicated on being able to cash out Darwin's Plan B by showing how simple possession and use of language could have given humans their cognitive powers. To do that requires us to find out how the kind of language capable of performing such a task could have evolved. So, inevitably, a great deal of this book has to be devoted to showing how language evolved.

Then some space has to be given to testing this account of language evolution. We can't (yet, at any rate) test it by examining prehistory, and can test it only indirectly by checking ancillary claims (e.g., in neurobiology). However, there is one way in which it can be directly tested. If it is correct, it must be able to give an account of how children acquire language that fits the facts of acquisition better than either nativist or empiricist accounts. But once all these tasks are completed, Wallace's problem, as Darwin's insight suggests, requires relatively little effort to solve.

The detailed organization of the book is as follows. Chapters 2 and 3 are occupied with some ground-clearing exercises, in linguistic and evolutionary theory respectively, since so much of what has been thought and written in these areas is misleading and if not attended to would simply get in the way of exposition. Chapters 4, 5, and 6 deal respectively and chronologically with each of the three processes outlined above. Chapters 7 and 8 test the model derived from earlier chapters by looking at how language is acquired by children under normal (Chapter 7) and abnormal (Chapter 8) conditions. In Chapter 9, we return to Wallace's problem and summarize its solution.

CHAPTER 2

Generative Theory

The theory of language that has most consistently upheld a belief in the existence of an innate component of language (and in particular, of syntax) is, of course, Chomskyan generative theory. When generative theory emerged in the late 1950s, it energized a hitherto lackluster field. Studies of syntax consisted of descriptions that were little more than labeled lists of constructions. There was no explanation of why language was the way it was, why it worked as it did. The distinction proposed by Chomsky (1957) between deep and surface structures (to be jettisoned forty years later, however) seemed, for generations raised on theories of a psychological "unconscious," to hint that linguistics stood on the brink of profound discoveries about the human mind. The revelation that most previous grammars were mere lists of exceptions and that the most interesting features of grammar might be the ones that grammars never mentioned (because they were shared by all languages, hence taken for granted) looked to mid-twentieth-century eyes like a novelty, although as Chomsky (1966) himself was quick to point out, it had conceptual roots in much seventeenth-century thinking. An eager willingness to board the generative bus and ride it to wherever it was going was, around 1960, perhaps the commonest reaction both among linguists and those in neighboring behavioral sciences, in particular psychology. Perhaps coincidentally and perhaps not, such feelings mirrored the prevailing optimism of those days, much as current cynicism and pessimism about the generative enterprise mirror the ominous forebodings that characterize the second decade of our current century.

Be that as it may, the generative bus did indeed travel fast and far. It is instructive to compare the bibliographies of generative writings in each of the succeeding decades. Hardly any names recur throughout; relatively few names persist through more than a single decade. As it went through its rapid but tortuous career, more and more travelers leaped from or were flung off the bus, unable to deal with apparent changes in direction and unable to understand what was wrong with the paradigm that was now being left behind. Interested bystanders were quick—much too quick—to assume that linguists simply couldn't make up their minds, that they were floundering in confusion, with no map to steer by.

In reality the changes in generative theory were simply the kinds of change you might expect to find in any vigorous developing science. If there was no map, it was because new territory was constantly being explored. Those in the forefront of discovery could do no more than follow their nose and make what seemed to be the best next move. Moreover there is no question that generative theory attracted some of the quickest and most incisive minds in science. It remains an indisputable if (for some) inconvenient fact that by far the greater part of what we know about syntax was discovered in the course of developing generative theory.

If we are to understand why generative grammar ultimately failed as an explanatory theory for linguistics and what it would take for such a theory to succeed, we need to understand how generative theory evolved over the years, why it made the decisions that it did, and how things might have gone differently if other decisions had been made. Accordingly there follows a brief history of the movement.

The Three Phases of Generative Grammar

Generative grammar has passed through a number of separately named avatars, but in reality its history falls into three clearly defined phases. Each phase developed along similar line. Each began with assumptions that seemed stunningly simple and straightforward at the time. However, as the years went by, each yielded more and more complex results. Eventually a point was reached where complications became top-heavy, and another set of seemingly straightforward assumptions replaced the first set.

While so brief a summary provides useful guideposts, detailed reality was much messier. For one thing, at each major shift some adherents of the previous phase would apostatize, some privately, some very publicly, and would either continue to work within the old paradigm or branch off in a new direction. For another, analytic practices that were not outright condemned by the new phase tended to persist into subsequent phases, whether or not they were appropriate there (and in Phase 3 even outright condemnation was not always enough to stop them). For a third, the transitions between phases were protracted; if we were to date the onset of Phase 2 from its earliest intimations, the transition between Phases 1 and 2 would become longer than Phase 1 itself. In what follows, dates are only approximate, and the end dates of phases are intended merely to mark the time by which the succeeding phase had become well-established.

Before beginning this account, however, one thing should be made clear: contra much criticism, apparent changes in the nature of generative grammar were thoroughly motivated, aimed at the same ultimate goal, and led to an overall increase in our understanding and knowledge of syntax. What appeared to many as missteps often arose, as Newmeyer (1993) pointed out, through the inevitable tension between descriptive and explanatory adequacy.

An ideal grammar would be both descriptively and explanatorily adequate; that is to say, it would combine full descriptive coverage of every syntactic structure in a given language with explanation (hopefully in terms of general principle rather than historical contingency) of why those structures were as they were and not otherwise. But this ideal may not be possible. In Sapir's (1921: 38) dictum, "All grammars leak." Grammars of different dialects of English would contradict each other. Even the grammars of individual speakers of the same dialect would often prove incompatible. A completely adequate descriptive grammar would be no more than a listing of all such discrepancies; like Borges's (1975) mythical map of the world, it would be as large (and as uninformative) as what it described.

From the beginning, generativists have been (as all scientific inquirers should be) more interested in explanation than description and have believed, in common with researchers in "harder" sciences, that explanation involved the discovery of generalizations that extended over a wide variety of superficially different cases. In other words, they believed that language, like other phenomena, was lawful. (If one didn't believe in the lawfulness of nature, what would be the point of science?) But to start such a program meant focusing first on what seemed most amenable to explanation and proceeding thereafter to tackle less tractable phenomena. Inevitably this involved at least temporarily losing empirical coverage of things that a more descriptive approach would have included. Many in linguistics and outside of it complained about this. But such critics simply showed that they didn't know how science worked.

Standard Theory (1957–1980)

Originally transformational-generative grammar (as it was first named) took the form of phrase-structure (PS) rules and transformations. PS rules were descriptions of basic single-clause structures in terms of abstract categories, such as S (sentence), VP (verb phrase), and NP (noun phrase). Normally these took the form of what were called "rewrite rules." For example, S—>NP VP meant that a sentence could be rewritten as (i.e., consisted of) a noun phrase followed by a verb phrase, and subsequent rules would similarly decompose NP and VP into their component parts until one reached the level of individual words. Transformations were processes that took basic PS rules and changed simple structures into more complex ones by transposing and/or replacing symbols and/or merging two simple structures into a single, more complex one. Thus there would be transformational rules for turning statements into equivalent questions, active sentences (*The police arrested Mary*) into equivalent passives (*Mary was arrested by the police*), dative sentences (*She gave the book to Bill*) into double-object sentences (*She gave Bill the book*), and two sentences with a common member (*The driver was Mary's brother* and *The driver met you*) into a single sentence with a relative clause (*The driver who met you was Mary's brother*).

The most controversial of the innovations that the new grammar required was the notion of movement. For example, a question such as *Who did Mary see?* was claimed to result from movement. *Who* was said to have originated in the object (postverbal) position in an "underlying" sentence (one that is actually found, note, in a "surprise" question), *Mary saw WHO?* A normal question was then formed by moving *who* to the beginning of the sentence.

People, even some linguists, took this literally. Of course it was a metaphor, a convenient shorthand for noting that the two possible positions of *who* were closely related in some way. The sentence-initial position was where (in English and many other, if far from all languages) question words must appear. But every nominal constituent of a sentence has to have some relationship to the verb—has to refer to who performed the action of the verb, who experienced it, what was used to perform it, or for whose benefit it was performed, and so on. Such relationships are called thematic roles (theta-roles for short) and include things like Agent (for the performer), Theme or Patient (for the performed upon), Instrument (for the thing used), and so on. In the sentence *Who did Mary see?, who* has a Theme/Patient role, and in English that role is always assigned to whatever noun or noun phrase immediately follows the verb. Moreover there were some very subtle effects suggesting that the "empty space" that "movement" supposedly left behind was in fact just as real as anything else. Did it ever occur to you, for instance, that while in casual speech it's okay to reduce *want to* to *wanna*, you can do this in *Who did you want to meet?* but not in *Who did you want to meet Bill?*

Since this is probably already more syntax than many readers want, I'll explain it in a postscript to the chapter. For now, the most important thing to note about the Standard Theory is that in its early years it was agnostic on issues of innateness. Its opening salvo, *Syntactic Structures* (Chomsky 1957), was hardly more than a how-to manual. However, the theory had obvious implications for how language was acquired and what might underlie it, implications explored in Chomsky (1965, as well as a number of subsequent works). The notion of an innate component crystallized well before the end of Phase 1, even though it remained short on detail.

The original impetus for hypothesizing an innate Universal Grammar sprang from the conjunction of two streams of thought, the union between which is first made fully explicit in Chomsky (1965). One stream goes back as far as Plato: the notion that experience cannot be the sole source of human knowledge. Another goes back at least to Humboldt: the rapidity with which children at an extremely early age master quite complex structures of human language. Plato's insights had been developed in a long tradition of rationalist philosophy, of which Chomsky perceived himself the heir. Humboldt's insights, on the other hand, had been largely ignored.

The Platonic connection alone practically ensured that Chomsky's union of these two streams should be couched in mentalistic terms. But one may well ask what prevented him, at a time when Lorenz and Tinbergen were developing complex theories of instinct (of which, incidentally, he was well aware), from at least considering the possibility that language was simply some form of instinctive behavior, taking the form of actual operations by the brain, rather than some form of tacit knowledge. In part Chomsky's failure to even consider this possibility may have been caused by the very word *behavior*, which had acquired connotations of Skinnerian antirationalism. But I think also that there may be another factor involved, a case of an unintended consequence, stemming from a mainly methodological decision that was made for what seemed, at the time, the best of reasons.

That decision was to make a sharp distinction between competence and performance. This distinction seems nowadays to have evaporated along with deep and surface structure and many other concepts that seemed for decades to form indispensable parts of generative grammar. Certainly the massive and comprehensive *MIT Encyclopedia of the Cognitive Sciences* (Wilson and Keil 2001) contains no entry on the distinction (although dozens of seemingly less salient aspects of generative grammar get their own articles). But in determining generative methodology, the competence-performance distinction played a crucial role.

If there was a single methodological innovation that made generative grammar possible, it was the realization that a serious grammar could not be created merely on the basis of sentences that had actually been uttered by native speakers. Previous treatments of sentence structure had assumed that such sentences formed the only evidence on which a grammar could be based. Leaving aside the more widely discussed issue of "performance error" (the fact that such evidence always contains sentence fragments, interruptions, slips of the tongue, and other data that one would not want to include in a grammar), any grammar so written could be no more than a descriptive list of permissible construction types, hence could not hope to achieve the goal of explanatory adequacy.

To meet that goal, it had to be possible to describe not only what speakers *could* say but also what they could not. Only by comparing sentences that native speakers accepted as grammatical with apparently similar sentences that they regarded as ungrammatical (conventionally marked by a preceding asterisk) could one arrive at even a first approximation of an answer to the *why* question. That answer was this: in creating sentences, speakers were following highly specific, albeit wholly implicit, rules. Such rules could best be understood by detailing the structures they could *not* generate as well as those they could and determining the factors, often quite subtle, that made them different. (See the postscript to this chapter.) When used for this purpose, the competence-performance distinction was a wholly beneficial, indeed indispensable tool.

Note that the existence of such implicit rules, though it might not in itself prove the existence of an innate Universal Grammar (UG), was fully congruent with it. Indeed the notion of UG provided an explanation without which the source of such rules was baffling. But the most beneficial distinctions can have unintended consequences. Chomsky had made competence rather than performance the major focus of inquiry. Indeed a theory of performance could hardly exist before one knew exactly what was being performed. But since the theory of competence was constantly widening and deepening, any theory of performance—performance of any kind, in any sense—was constantly being postponed.

Postponing the study of performance meant that no one ever had to say exactly what was meant by it. Performance certainly extended over language use, the circumstances under which and the purposes for which actual sentences were produced on actual occasions. Language use was excluded, very rightly, from consideration. But what about the series of actions the brain had to perform every time a sentence was uttered? Was that performance? Given almost total ignorance of how the brain produced language, there seemed little point in even considering such issues. Yet, as one generativist observed at the time, "the linguistic description and the procedures of sentence production and recognition *must correspond to independent mechanisms in the brain*" (Katz 1964: 133, emphasis added).

Unfortunately nobody seems to have thought to ask the follow-up question: If independent mechanisms recognize and produce sentences, what is the point of hypothesizing a higher layer of abstract knowledge? What would that layer do that was not already being done? Recall, however, that in the 1960s and 1970s focus was on psychological rather than neurological reality, on the mind rather than on the brain, and on cognitive rather than electrochemical operations (although it was beginning to be realized that these pairs of names might be referring to different aspects of the same things). The reason for this was undoubtedly the contemporary state of brain-language studies, which can be fully appreciated only by reading a contemporaneous review article such as Brain (1961); there are long paragraphs on speech mechanisms and aphasia, but the only reference to syntax is a brief mention of "the difficult problem of serial order" (437), with no attempt to solve or even state the problem. Before brain imaging, the idea of a neural flowchart for real-time sentence construction was inconceivable; "psychological reality" was the only game in town.

The most serious attempt to link generative grammar with actual neural operations was the derivational theory of complexity (Miller 1962). This theory made a simple and seemingly sensible initial proposal. If grammatical transformations (the mainstay of the first phase) represented actual operations that individuals had to carry out in producing and comprehending sentences, surely sentences with more transformations should be more difficult to produce than sentences with fewer transformations. Hence, other things being equal, sentences with more transformations should take longer to process and should appear later in acquisition. This was an eminently testable proposition. Unfortunately "tack[ing] a psychological tail to a syntactic kite" (Miller 1975: 204) turned out to produce, at best, equivocal results (Fodor and Garrett 1967; Brown and Hanlon 1970)—hardly surprising in view of the fact that the transformations themselves would shortly be abandoned by generativists. But the failure of this approach helped to turn researchers away from any attempts at integration with neighboring disciplines. From then on, they would not "take account of any data other than primary linguistic intuitions" (Pylyshyn 1973: 437).

In the first phase of generative grammar, the consequences of this self-denying ordinance were not serious. It seemed reasonable enough to argue that what was known about the brain was too little and too conjectural to influence the nature of grammars. Moreover during a large part of that first phase, the assumptions of grammarians and the structures of the grammars themselves were changing too rapidly and too extensively for anyone to say exactly what kind of knowledge grammars

24 Generative Theory

should represent. However, the ultimate nature of the innate component seemed already to be clear. Chomsky (1968: 65) stated that "the real problem for tomorrow is that of discovering an assumption regarding innate structure that is sufficiently rich, not that of finding one that is simple or elementary enough to be 'plausible.'" And again he left no doubt that in any neurological research, the neurologist must play second fiddle, unquestioningly accepting the guidance of the linguist: "The discoveries of the linguist-psychologist set the stage for further inquiry into underlying mechanisms, inquiry that must proceed blindly, without knowing what it is looking for, in the absence of such understanding, expressed at an abstract level" (Chomsky 1988: 7). The linguistic cart would pull the neurological horse, even though, in real life, it was the horse that was doing the real work, and language was just part of the work it did.

But even as all this was happening, the theory was in trouble. There had to be more rules than anyone had expected, dozens or even hundreds just for English. Moreover these rules had to apply in a determinate order, since the output of rule X might have to form part of the input to rule Y. Unfortunately it often happened that there were unimpeachable reasons for ordering X before Y, but equally unimpeachable reasons for ordering Y before X. And rule proliferation and "ordering paradoxes" were only two of a number of problems that led to the eventual replacement of the Standard Theory.

Principles and Parameters (1981–1994)

The theory that finally emerged from the long transition between Phases 1 and 2 formed a marked shift in the direction of putting explanatory adequacy above descriptive adequacy and had its birth in a seminal dissertation (Ross 1967) proposing a handful of general principles that would subsume many particular, construction-bound rules. After agonizing birth pangs (described in Harris 1993) there eventually emerged a tightly meshed and comprehensive theory that divided syntax into a half-dozen subtheorems dealing with such things as Binding (determining the referents of items like pronouns that, unlike proper names, didn't have reference built in) or Theta Theory (dealing with the thematic roles of sentence constituents: Agent, Theme or Patient, Instrument, etc.). Principles of the subtheorems were supposed to apply across the board (i.e., were exceptionless) and across all languages (i.e., were universal) and were of an extremely abstract nature. For instance, the Empty Category Principle stated that all positions from which an item had been moved—such as that indicated by *e* in *What does John want e,* a sentence supposedly derived from *John wants what*—had to be properly governed. To understand what this means involves understanding what is meant by government, which in turn involves understanding a structural relation known as c-command. This will suggest the extent to which argumentation in syntactic studies had already become theory-internal—driven by developments in the theory itself rather than by details of empirical data. This is by no means necessarily a bad thing per se—what we find in subatomic physics is far indeed from the "reality" we think we perceive—but it carries with it a number of dangers, one of which is that assumptions may be based on assumptions based on yet other assumptions, one or more of which may rest on inadequate evidence.

Principles took care of supposedly universal features, but left many things that were variable across languages. What was to be done about them? To understand the nature of the problem generativists faced at this stage, we need to look at a potential ambiguity in the word *universal.*

One obvious sense of *universal grammar* is a grammar that, because it is species-specific and hence presumably forms part of the human genome, is shared by all humans regardless of whatever language they may come to speak. But there is another potential sense of "universal" here. If such a grammar is to be shared by speakers of all languages, and if indeed it is to be a means of acquiring language—without which, presumably, no language at all can be acquired—then UG has to be equally capable of generating any of the world's thousands of languages. Consequently UG would somehow have to account for the ways languages varied from each other, as well as ways they can be shown to be the same. It must be universal in the sense that it must account for any and all forms of grammar found in human language. There is a brief mention in the introduction to Chomsky (1981) of a distinction between "core" grammar (which UG must account for) and a "periphery" of idiosyncratic constructions (which presumably it need not). But no such distinction has ever been formalized or even explicitly stated. For practical purposes, constructions that varied cross-linguistically have generally been treated as if they formed part of core grammar.

The means chosen to deal with variable constructions consisted of what were described as "parameters"—ranges of possible variation at particular points in the structure of sentences. Children could then set a given parameter on the basis of experience; in a best-case scenario, parameters, like light switches, would come with only two possible settings. Take the so-called Pro-Drop or Null-Subject parameter. In English and similar languages, subject pronouns can't be dropped; you have to put them in even if you can't think what they might refer to, like the *it* in *It is raining*. In Spanish you'd say just *Llueve,* literally "Rains." If you put in a pronoun, **il llueve,* any Spanish speaker would look at you as if you were crazy, and indeed Spanish subject pronouns seldom appear anywhere unless they are required for emphasis or disambiguation. Sometimes the parameter would be a simple matter of directionality, such as the Head-Modifier parameter, which determines whether a head (noun or verb, usually) precedes or follows its modifier.

Concomitant with these developments, the transformational component, which had loomed so large in the Standard Theory, had shrunk to a single rule, "Move alpha," meaning move anything anywhere. But this apparent recipe for syntactic anarchy was in practice stringently constrained by the various principles in such a way that movement anywhere was strictly limited. In any case, conceptual leanness in the transformational component was more than offset by conceptual proliferation in the other components. Overall, as compared with the Standard Theory, the innate component within principles and parameters (P&P) had grown enormously in size, in complexity, and in specificity due to Chomsky's (1986: 55) conviction that "the available devices must be rich and diverse enough to deal with the phenomena exhibited in the possible human languages."

The notion of parameters and their role in syntax looked like a bonanza to students of language acquisition, and one result was an explosion of acquisition studies within a P&P framework (e.g., Grimshaw 1981; Valian 1981, 1986; White 1985; Hyams 1986). Researchers now had far more specific proposals to test, and they could test them from a comparative perspective in a variety of dissimilar languages. Indeed such was its attraction to acquisitionists that the principles-and-parameters approach still persisted in acquisition studies more than a decade and a half after the initial emergence of Phase 3, with its very different assumptions about language—assumptions that, on at least one view, should have rendered such studies illegitimate (Longa and Lorenzo 2008). Yet thirty years after P&P's original formulation, there is still not even a tentative list of possible parameters, nor even any idea as to how many parameters there are. After twenty years Baker (2001) estimated "10 to 20." Four years later Roberts and Holmberg (2005) put the probable figure at "50–100," an estimate they described as "conservative."

With prescient insight, Newmeyer (1996: 64–65) observed, "In the worst-case scenario, an investigation of the properties of hundreds of languages around the world deepens the amount of parametric variation postulated among languages, and the number of possible settings for each parameter could grow so large that the term 'parameter' could end up being nothing more than jargon for language-particular rule." This is exactly what happened over the next decade or so. Many already proposed parameters turned out to defy the limitation to binary settings. In many languages, the position of heads relative to their modifiers varied according to the word classes these belonged to; in English, to give just one example, adjectives precede nouns while noun-modifying prepositional phrases follow them. As more and more parameters had to be hypothesized, more settings had to be permitted, and more counterexamples turned up. Even before Newmeyer's prophecy, some of these were apparent (Gilligan 1987); subsequently more and more emerged (among others, see Newmeyer 2004, 2005; Haspelmath 2008; McWhorter 2008). In the past couple of years, the idea of parameters has been abandoned even by figures as central to generative grammar as Boeckx (2010).

But quite apart from any empirical shortcomings, the load being placed on UG was becoming extraordinarily heavy in kind as well as in quantity. For instance, one had to assume, as just one small part of that knowledge, "binding theory." Binding theory involved the relationship between different types of pronouns and their referents and explained (among many similar things) why, if you said *Susan washed her*, *her* couldn't refer to Susan, but if you said *Her sister washed Susan*, it could, even though in *Susan made her sister wash herself*, *herself* couldn't refer to

Susan but must refer to her sister. According to its original formulation (Chomsky 1981: 184–185), a constituent cannot be bound (that is, have its reference determined) outside its "governing category." And to understand "governing category" you had to understand two more abstractions, "minimal category" and "governor," and all these abstractions had to be taken into consideration before one could even determine how any of the three "binding principles" applied in any given case. Knowledge of all these conditions, categories, and principles represented only one small part of what, under P&P, one had to attribute to the innate component and hence to the minds of prelinguistic children.

If anyone had asked Chomsky or any other generativist exactly how knowledge of this degree of abstraction and language specificity could be instantiated in the mind/brain, his answer would presumably have been along the lines of Chomsky (1986: 26): "The steady state of knowledge (I-language) attained and the initial state ($S0$) are real elements of particular mind/brains, aspects of the *physical world*, where we understand mental states and representations to be *physically encoded* in some manner" (emphasis in original). In other words, they (somehow) must form part of brain operations, though exactly how was left for neurologists to figure out.

P&P made an excellent discovery procedure, revealing many things about language that any future theory would have to account for. Yet even as it flourished, Chomsky himself was about to reject many of the assumptions on which it was based.

The Minimalist Program (1995–Present)

It should be emphasized that, as Chomsky himself has repeatedly stated, the Minimalist Program (MP) is just that—a program, a way of approaching language, not an explicit theory—and that it might be realized in a variety of different ways. Its motivating force was Chomsky's (1995: 233) realization that the assumptions of earlier versions of generative grammar were all too often "of roughly the order of complexity of what is to be explained." The answer was to reduce grammar to "virtual conceptual necessity," the minimum number of elements that might suffice for the derivation of sentences. That number was spelled out as "elements already present in the lexical items" selected for a given sentence, to which "no new objects are added . . . apart from rearrangement of lexical properties" (228).

But most of the underlying principles and processes of the P&P framework fell outside of these elements. Among these were "D-Structure; S-Structure; government; the Projection Principle and the Theta Criterion; other conditions held to apply at D- and S-Structure; the Empty Category Principle; X-bar theory generally; the operation Move Alpha; the Split-I hypothesis; and others" (Chomsky 1995: 10) as well as "Binding Theory, Case Theory, the Chain Condition, and so on. . . . There should be no government, no stipulated properties of chains, no binding relations internal to language, no interactions of other kinds . . . no phrasal categories or bar levels, hence no X-bar theory or other theory of phrase structure apart from bare phrase structure" (Chomsky 2007: 4–5). Properties that could not be deduced from "conceptually necessary" ones would be eliminated altogether. All that remained of syntax proper consisted of only two components. One was Merge, a process that takes any two syntactic objects (words, phrases, clauses, etc.) and joins them to form a new syntactic object. All that stopped anything from merging with anything else was the necessity for the output of the process to be interpretable at the two "interfaces" with the other necessary components of language: sound and meaning (or in Chomsky's own terminology, SM, the Sensory-Motor system, and CI, the Conceptual-Intentional system, previously known as Logical Form). The only other component was Move (already discussed), and hopefully Move could be merged with Merge by treating it as Re-Merge, merging a copy of something that had already been merged once.

In terms of its conception, the MP was dazzling. In terms of its effects, it was bewildering. The license given by the fact that the MP was a program that could in principle be developed in a variety of ways allowed syntacticians to retain features of P&P and earlier versions if these seemed indispensible. But while at least some syntactic analyses tried to adhere to the basic intentions of the MP, its impact on studies of acquisition and evolution was (no pun intended) minimal.

Generativists studying language acquisition have largely ignored the MP and continued to use a parameter-setting framework, even though parameters had no defined role to play in the new framework (Longa and Lorenzo 2008). Conversely, while it has been suggested (Golumbia 2010) that evolutionary concerns may have motivated the MP, the evolutionary proposals that issued from it (Hauser et al. 2002; Chomsky 2010) managed at the same time to alienate both linguists and biologists—linguists by an excessive focus on alingual species, biologists by making assumptions that few if any experts in evolutionary biology could accept. Even in syntax results have been equivocal. One linguist, who felt free to give in a course handout an opinion he might have hesitated to profess in a more public venue, confessed with rare candor, "Practically speaking, what happened was a change in the fundamental perspective on what is happening in syntax, but it turned out to have little effect on the day-to-day life of syntacticians" (Hagstrom 2001).

It is therefore hardly surprising that the implications of the MP for the nature and content of UG have been largely ignored. But those implications are surely dramatic. All of the heavy baggage the component had to carry, such as the principles of binding mentioned earlier, has now disappeared. It is possible to reduce binding, for instance, to movement (Hornstein 2001), and movement may be reduced to Merge (Koster 2007). And since Merge is hypothesized as possibly the only operation unique to language (Hauser et al. 2002)—that is, not developed for or utilized by other cognitive systems—it becomes potentially the sole content of the innate component for language.

So much for principles. What about parameters? In the previous section we saw growing skepticism among generativists with regard to parameters. But in any case, since Borer (1984) it has been widely assumed that parameters are located in the lexicon, particularly in grammatical morphology ("function words" and affixes). And morphology has to be learned, including the functions of the various grammatical morphemes. If you think this is straightforward, consider "simple" locative prepositions like *in, at, on,* and so forth and their "equivalents" in Spanish or Dutch (to go no farther than Europe); there are few if any pure matches, and the functions of an item in one language may be wholly or partially divided between two or more items in the other, or vice versa. That all such idiosyncratic properties of particular languages are somehow preprogrammed in the child seems inherently unlikely. At most there might be a semantic set of possible locations (above-below, inside-outside, etc.) that the child then has to match up with whatever set of locative prepositions the target language offers—an onerous task, but not one beyond the power of general learning tools.

In a decades-long battle against empiricism that began with his demolition of behaviorist linguistics (Chomsky 1959), Chomsky had long insisted that the innate equipment for language must be task-specific and wholly distinct from the general-purpose mechanisms that, according to empiricists, were what enabled children to learn language. Now, without any explicit admission, he seemed to have abandoned most of these claims. Could it be that there was now nothing (apart from Merge) to distinguish generativism from empiricism?

Minimalism and Empiricism

Although the MP has been around for a couple of decades, the cognitive science community has yet to grapple with some of its broader implications. I have found only three discussions of the relationship between the MP and empiricism, all quite recent, one by a longtime generative linguist (Hornstein 2009), the other two by relatively little-known scholars (Collins 2010; Golumbia 2010).

Collins (2010: 2) cites Chomsky (1965: 58) as saying, "The empiricist effort to show how the assumptions about a language acquisition device can *be reduced to a conceptual minimum* is quite misplaced" (Collins's emphasis). Since "reducing" the language acquisition device (which is equivalent to UG or the innate component) "to a conceptual minimum" is the main goal of the MP, Collins (2010: 5) concludes, "Bizarrely, on the face of it, minimalism is an empiricist agenda."

Collins attempts to exonerate Chomsky from this charge by arguing that while the part of the innate component devoted exclusively to language may have shrunk, language itself remains a complex and at least partially innate faculty. Moreover the "poverty of the stimulus"—the fact that linguistic data are inadequate for a child to arrive at a grammar—applies no matter whether the child's equipment is richer or poorer, more or less specific to language. Similarly Hornstein (2009) claims that Chomsky's goal was never to fulfill some nativist agenda but rather to seek a solution to "Plato's problem"—the inadequacy of experience to account for all knowledge—wherever one might be found, and let the chips fall where they may. Even if most or even all of language in its broad sense was derived from faculties that had evolved for other purposes, these faculties were still innate, and the mind/brain was anything but the tabula rasa that empiricists supposed.

But these arguments miss the point. It was specifically the complexity and the innateness of task-specific mechanisms for language, and the dissociation of these from other cognitive faculties, that had both formed the substance of Chomsky's discourse for decades and served as the main target for empiricist attacks on him. Indeed years after the MP had emerged, Chomsky (2002: 85–86) was continuing to claim, "It is hard to avoid the conclusion that a part of the human biological endowment is a specialized 'language organ,' the faculty of language (FL). Its initial state is an expression of the genes, comparable to the initial state of the human visual system. . . . As discussion of language acquisition becomes more substantive, it moves to assumptions about the language organ that are *more rich and domain-specific,* without exception to my knowledge" (emphasis added). This suggests a disconnect between Chomsky's syntactic practices and his metatheoretical assumptions that has been neither resolved nor explained.

The other paper in question, Golumbia (2010), never mentions the e-word, accusing Chomsky of functionalism rather than empiricism. But this is clearly a mistake. Functionalists share with empiricists a distaste for formal analysis and a strong skepticism about any kind of dedicated innate machinery for language. But they have at least one belief that empiricists do not necessarily share and that Chomsky continues to strongly reject: they believe language evolved for communicative purposes and is shaped throughout by communicative needs. The fact that Golumbia lists Putnam and Quine—two empiricists with no commitment to functional linguistics—as two of the thinkers whose position Chomsky is now seen as approaching serves to further confirm that empiricism rather than functionalism is what he too is really talking about. One can only conclude that, while Chomsky's position is still some distance from full-on empiricism, he has (intentionally or otherwise) substantially narrowed the gap between empiricism and nativism.

Minimalism and Biology

Was the creation of the MP motivated, at least in part, by Chomsky's realization that the complex P&P model of the innate component was fundamentally incompatible with evolutionary biology? Golumbia suggests that this is the case. But he also makes an even more penetrating observation: "Under MP it seems misguided to call UG a 'system.' It is instead a set of operations" (Golumbia 2010: 40). In other words, and without any intention on Chomsky's part, the MP has brought syntax measurably closer to the proposals argued for in this book.

Here the way UG is conceptualized is key. From the beginning, and as expressed in the title of one of his most influential books (Chomsky 1986), Chomsky has conceived of UG as representing "knowledge." This upset many philosophers for whom knowledge was something you knew you knew, and Chomsky (1980: 69) was obliged to coin the ugly neologism *cognize* to characterize the kind of knowing that gave rise to UG. However, "someone who cognizes cannot tell what he cognizes, cannot display the object of his cognizing, does not recognize what he cognizes when told, never forgets what he cognizes (but never remembers it either), has never learned it, and could not teach it. Apart from that, cognizing is just like knowing!" (Bennett and Hackert 2007: 138).

A more perspicuous way of conceptualizing what UG is would indeed be "a set of operations." After all, and regardless of whether or not the mechanisms involved are innate, the brain has to perform a series of operations in order to produce sentences. Consequently there is a redundancy if the brain contains both autonomous neurological mechanisms for generating sentences and a layer of abstract "knowledge," a grammar, with a content equivalent to that of those mechanisms One or the other is surely superfluous, and since it is not possible to eliminate the physical operations, we should abandon the abstract knowledge. By so doing—by conceiving UG as the result of a set of operations, routines, or algorithms that the brain performs—we align language with a host of other processes (e.g., digestion, circulation of blood, vision) of whose specific modes of operation we are subjectively just as unconscious.

It is worth taking a moment to see how Chomsky's current assumptions about syntactic operations could indeed serve as an initial hypothesis about how the brain actually constructs sentences.

Minimalist Operations

We need first to understand the difference between two different approaches to syntactic description: the representational and the derivational. Representational approaches are top-down approaches. All forms of generative grammar that preceded the MP were representational. They began with fully fashioned sentences, broke them down into their component parts (phrases and clauses), and then broke these down into *their* component parts (word stems and bound morphemes, a morpheme being the smallest independent unit of meaning and a bound morpheme a suffix or prefix like *un-* or *-ing* that has to find a word stem to attach to). Typically sentences were displayed in terms of hierarchical tree structures; a complete structure had to be assembled before words could be inserted into it.

With the emergence of the MP came Bare Phrase Structure (Chomsky 1994), a bottom-up procedure that constituted a precise mirror image of the representational approach. Two words would be combined via Merge, a third combined with the unit created by the first Merge, and the process would be repeated until the structure was complete. The representational (Brody 2002) and the derivational (Epstein et al. 1998; Seely 2006) approaches still coexist, albeit uneasily, under the broad umbrella of the MP. Because of this mirror-image relationship, and because the end product of a derivation by Merge is a hierarchical tree structure similar to those from which the representational approach starts, one might conclude that the two are simply notational variants and that it makes little practical difference which one chooses.

That is indeed the case if one's sole concern is to express the content of some form of "knowledge." If, on the other hand, the purpose is to characterize operations that a brain might actually carry out, then starting from the bottom up provides a far more helpful model than trying to start from the finished product. Let us therefore go through the steps of a derivational representation and see how far the analogy will hold. The operation begins with the Numeration (Chomsky 1995: 225–228), which assembles all the words to be used in the sentence, with the number of times each word is to be used. Clearly the brain must (in some sense) perform an essentially similar operation. In reality items are probably retrieved over time rather than all at once, but here the "idealization of instantaneity" is innocuous.

The process proceeds by assembling words into structures: merging two words to form a phrase, then a third to enlarge that phrase, and so on. Clearly, although the ordering of each pairwise juncture will eventually correspond to that in the ultimate linear form of the sentence, a sentence will not be constructed sequentially in its final linear ordering (*The girl . . . The girl watched . . . The girl watched her . . . The girl watched her dog*). Spatial contiguity is not what determines the closeness of relationships between words. Take a sentence like *The letter I sent you should have answered that question*. The words *should* and *sent* are both directly adjacent to *you*, and *You should have answered that question* is in itself a fully grammatical sentence, where *you* and *should* are indeed closely related. But this sentence has a meaning completely unrelated to that of the previous one. In the previous sentence, although *you* and *should* are equally adjacent, *you* has a close relationship with *sent* (it is the object of that verb) and only the most indirect relationship with *should*, whose subject is not *you* but *the letter I sent you*. In other words, what are first merged are the words that are most closely related to one another, regardless of their ultimate linear sequence. Consequently, in our original sentence, *her* will first merge with *dog*, then *watched* with *her dog*, then *the* with *girl*, and finally *the girl* with *watched her dog*.

The processes just described may not always continue uninterrupted until the entire sentence has been constructed. Chomsky (1999) has hypothesized that derivations proceed by "phases." This article is so technical that an explanatory gloss on it, longer than the article itself, had to be written (Uriagereka 1999). Accordingly make the (extremely oversimplified) assumption that a phase is roughly equivalent to a clause. Chomsky suggests that phases, once complete, become impenetrable; no further syntactic processes can apply to them (except, of course, mergers with other phases to yield complex, multiclausal sentences). This would help to account for the widespread "island effects" found generally in syntax, explaining, among other things, why (except under some highly specific circumstances) constituents in general cannot "move" outside of the island-like clauses into which they were originally merged.

Again we find similarity between a derivational approach to syntax and a process that the brain itself quite likely uses. Chomsky never discusses whether he sees phases as being produced sequentially or simultaneously, but there is no need to rule out simultaneous processing and there are some good reasons for supposing it. It is widely accepted that the human brain is a massive parallel processor (Ballard 1986; Alexander and Crutcher 1990; Wassle 2004), so simultaneous production of phases is certainly possible, and the speed of speech seems to render it likely. Speakers can maintain speeds of 150 to 180 words per minute in normal conversation. While there are no reliable estimates of average sentence lengths (something that probably varies widely across genres), an average of around fifteen words is often suggested, though that figure is surely less in casual conversation. But even a fifteen-word average would yield a rate of ten to twelve sentences per minute, or only five to six seconds per sentence. Given that this time period must include word retrieval and sentence assembly as well as utterance, it seems unlikely that the brain could maintain such rates if it did not assemble parts (phases?) of longer sentences simultaneously and in parallel.

In contrast with the derivational model, mechanisms similar to those of a representational syntax seem relatively unlikely to be used by the brain. If they existed, they would have to be capable of first creating an abstract tree for every sentence, a large and complex one in many cases. Only once such a tree, with all its nodes and branches, had been constructed down to the level of terminal nodes would it be possible to retrieve and insert the necessary lexical items. This would amount to planning the structure of a sentence in detail before one even knew what one wanted to say. Such is our ignorance of the brain that we cannot rule out such a procedure (or indeed virtually *any* procedure) entirely, but it is hardly the first possibility one would want to consider, especially given that a much more tractable one is available.

The similarities between a derivational version of the MP and putative workings of the brain are more likely coincidental than intentional. Chomsky expresses no awareness of any such connection and has never taken as an explicit goal the matching of known syntactic processes with hypothetically possible neural ones. But even if the degree of similarity is serendipitous, it is still there for more explicitly process-oriented proposals to build upon.

Chomsky's Problem

We have now surveyed a half-century of attempts to solve "Plato's problem," at least insofar as that problem affects language. We should now be able to get a better idea of "Chomsky's problem," which was how to produce a solution for Plato's problem that would align the study of language with other branches of scientific inquiry.

One cannot repeat too often that, despite all the criticisms (including outright abuse) that have been heaped upon it, the generative movement has had profoundly beneficial effects on our understanding of language. It is equally true that the shifts in generative theory do not represent any uncertainty about the goals it was pursuing but more closely resemble the behavior of some craftsman, a plumber, say, who in attempting to loosen a recalcitrant joint periodically shifts his grip on the pipe. As Hornstein (2009) and Boeckx (forthcoming) suggest, the devious developmental course followed by generative theory may have been the only practicable way to uncover all the linguistic facts that a final model of UG would have to account for. Yet for all the success it has achieved, generative linguistics remains curiously isolated from other scientific fields. (Indeed many would even deny it the status of a science.)

In part this state of affairs is due to factors that are no one's (or everyone's!) fault. The attempts of linguists to grapple with syntactic complexities have given rise to an arcana of opacity, a plethora of technical terms guaranteed to turn off all but the most dedicated inquirer. While material fully as complex as that of linguistics is routinely taught in high school physics courses, few people have any real exposure to linguistics outside of graduate school. Consequently nonlinguists have been reluctant to do the homework necessary for any meaningful dialogue with linguists. Naturally this in turn has encouraged a "take it or leave it" attitude among linguists that has colored attempts to remedy their isolation.

What Jenkins (2000) described as the "unification problem" has always been clear to Chomsky (1991: 6), who framed it thus: "How can we integrate answers to these questions [what "knowledge of language" is, how it evolved and is acquired, used and instantiated in the brain] within the existing natural sciences, *perhaps by modifying them?*" (emphasis added). But far from modifying linguistic answers, or even discussing how they might best be integrated with the findings of other sciences, Jenkins (2000: 3) indignantly rejects any talk of the "psychological reality" or "neurological reality" of linguistic hypotheses, declaring that "linguistics could now suggest core internal properties of the language faculty, that in turn posed important questions for biology," and that "the syntactic computations of the language faculty *are* the biological evidence" (emphasis added).

The standard argument in favor of this approach is concisely summarized by Smith (2000: viii): "The assertion that the physical or the physiological has some kind of priority is misconceived: theories in linguistics are as rich and make as specific predictions across a wide domain as do theories of chemistry or biology. Trying to reduce linguistics to neurology in the current state of our understanding is then unlikely to be productive." But this argument misses the point. No one is even suggesting that "the physical or the physiological" should have priority over linguistic theories, merely that the latter should not proceed as if the former didn't exist. We may not know much about the brain, but we know that when it comes to the structure of sentences, a derivational model is much closer to the kind of thing brains routinely do than is a representational model. We may not have a perfect understanding of evolution, but we know that if there is any kind of UG, it must have evolved, and we know that much of what has been proposed for UG is extremely unlikely to have been produced by evolution.

With the wisdom of hindsight it can be seen where generativists went wrong. From early in its career, generative grammar suffered from a deep internal incompatibility between its novel and its traditional elements. On the side of novelty, the faculty of language was conceived of as a "language organ," something like lungs, heart, or stomach, that operated, like those other bodily organs, at a level opaque to the rational, conscious mind. Language in the individual was described as "developing" or "growing," like pubic hair or wisdom teeth, rather than "being learned" or "being acquired." Yet on the other hand, language had traditionally been regarded as part of "human cognition," and generativists persisted in so treating it, hence in characterizing the content of UG as "knowledge of language." Curiously this incompatibility seems rarely if ever to have been noticed by generativists themselves.

Certainly its implications have never been fully discussed or analyzed at length.

Language has always been central in what came to be known as cognitive science. Perhaps we should consider whether language, at least as far as its basic syntactic structure is concerned, lies outside of cognition, which it unquestionably enables but of which it does not, in and of itself, constitute a part. Language would then have two clearly dissociable aspects: an inbuilt, task-specific mechanism for producing basic syntactic structures and a task-general mechanism (or mechanisms) for learning words and the lower-level syntactic processes for assembling them (not limited to those involving linear order) in ways that would enhance the value of language as a means of communication.

Abstract as such a hypothesis might seem, it can be easily tested. If it is correct, it should be possible to develop a theory of language that will achieve two distinct goals. The first goal is to produce a model of language that will be plausible in terms of biology, neurobiology, and the behavioral sciences in general. The second is to produce a model that will have explanatory power when applied to the various dynamic processes that language has undergone or continues to undergo. In other words, it will predict and necessitate features of language evolution, language acquisition, creolization, and linguistic change that might previously have been regarded as arbitrary or as produced by diverse causes. To advance such a theory is the main purpose of this book.

The logic of the MP has forced generative grammar in the direction of the model proposed in this book. In a striking anticipation of its general outline, Boeckx (2010: 26) wrote, "Once UG is seen to be much more underspecified than we thought, the very existence of variation receives a straightforward rationale: there is variation precisely because the genome does not fix all the details of Universal Grammar. . . . A Minimalist view of language makes variation inevitable." In the pages that follow, this model will be made explicit; Chapter 5 will describe the content of the "underspecified" UG, and Chapter 6 will describe the variation that this underspecification imposes on language.

But the model represents more than the logical goal of the generative enterprise. It is also the model of language that is most consistent with all we understand so far about evolutionary biology. The next two chapters show why this is so.

Postscript

I promised that at the end of this chapter I would further discuss why we cannot reduce *want to* to *wanna* in sentences with certain properties. Consider the following sentences:

1. Who do you want to (wanna) meet?
2. I want to (wanna) meet Mary.
3. Who do you want to (*wanna) meet Mary?
4. I want John to meet Mary.

(2) is a grammatical (if long-winded) answer to (1), and (4) is a similar answer to (3). Note that in (4), *want to* cannot be reduced to *wanna* because *John* intervenes between *want* and *to*. But in (3) there does not appear to be anything between them. However, that intervening position is exactly where *who* would have occurred in the "surprise" or "confirmation" question (5).

5. You want *who* to meet Mary?

This is the position that would normally be occupied by the subject of the subordinate clause SUBJECT to meet Mary. Merge must first of all locate all constituents in their "home" clauses; only subsequently can they actually appear at the beginning of the sentence. If we adopt what has been termed "the copy theory of movement" (another feature that seems to bring theory closer to what a brain might actually do), then a copy of the original *who* is what appears in sentences like (1) or (3). Since *who* is the subject of *meet* it must originally merge in the same clause as that verb. Since *who* is also something that is being questioned, it must (at least in English) appear in initial position. So the brain first assembles (3) with *who* in the position it occupies in (5), then merges the copy, so that what the brain constructs is actually (6):

6. Who do you want who to meet Mary.

The second occurrence of *who* is simply not pronounced, since there are restrictions (too complex for discussion here) that in certain circumstances prohibit the reoccurrence of particular constituents. But because something isn't pronounced doesn't mean that it isn't there, in some form substantial enough to block the *wanna* contraction. Empiricists who find this distasteful should think of the zero plural form for *sheep*, which allows us to say *one sheep is* . . . but not *two sheep is* . . . In general, it is the failure of empiricists to take account of linguistic facts as specific and subtle as those of *wanna* contraction that has prevented generativists from taking empiricist arguments seriously.

CHAPTER 3

The "Specialness" of Humans

The model of language proposed here is evolvable, and alternative models are not. That is a very strong claim, but strong claims are invariably better than weak ones. To refute a weak claim doesn't take you very far into the subject. To refute a strong claim involves digging much deeper, which is what the field badly needs. Too many unsupported assumptions have been made, and too many misunderstandings have been allowed to pass unchallenged. None of these misunderstandings carries more danger than those that involve the relations between humans and other animals. Some of those who probably regard themselves as staunch supporters of evolutionary continuity between nonhumans and humans behave as if normal evolutionary assumptions and approaches do not apply when dealing with humans.

The one good thing about behaviorism was that it treated humans and other animals exactly alike. All were subject to the same simple mechanisms of stimulus, response, and reinforcement. Unfortunately behaviorists ignored the fact that all animals are different, not in their rankings on some scale but in their distribution across an ecological landscape. No two species can occupy exactly the same space in that landscape; each species is constrained in what it can do by the exigencies of its niche. Cognitivists were ready to accept this. But even more important to them were the interactions between behavior and mental processes. And when it came to mental processes humans seemed to be in a class of their own.

For many, perhaps most, cognition came to mean primarily the kinds of conscious, voluntary mental processing that probably only humans can do. It is not altogether clear why this should be so. Mental processes mainly involve the processing of information, something that every animal with a brain does on a daily basis. A bat hunting insects at night, for instance, carries out complex computations over a rapid and constantly varying stream of information. Moreover this is not a stream independent of the bat. The bat deliberately produces the stream by generating sound impulses that can reach rates as high as 200 per second. Complex neural devices are required to decipher and interpret the echoes that these impulses create (Griffin and Galambos 1941; Griffin 1959). Griffin and Galambos first reported on their findings at a time when human sonar and radar research were still highly classified military secrets, and the initial reaction of the academic community was one of frank incredulity (Dawkins 1986: 35). If echolocation was something that required all the sophisticated ingenuity humans could muster, why should not this, just as much as language, be regarded as some form of cognition? Yet the phrase *bat cognition* is very seldom seen in work on bat behavior. (If you enter it in Google Scholar, the first work that comes up is about baseball.)

Consider organisms of a different kind that also show a unique behavior, but one that depends not on rapid processing of information but on a form of activity that (like language) would seem to require some kind of internalized plan, a neural template flexible enough to be adapted to current conditions. Take as an example the orb web–spinning spider, which first makes a bridge between two branches (or other convenient objects), forms a V shape below that, converts the V into a Y, links the points of the Y to create a frame, lays a large number of radius threads linking the frame to the center of the web, fills the frame with a spiral figure, then uses that spiral as a reference to lay a second spiral of sticky thread, finishing with a signal line that will alert it if an insect becomes entangled in the web (Harris n.d.). This is not a case where all members of the same species blindly generate an identical pattern. To the contrary, there has been "long-standing and repeated documentation of substantial *intra*specific variation in at least gross web characters such as number of radii, spiral loops, spacing between loops, angle of web plane with vertical, web area, top-bottom asymmetry, and stabilimenta" (Eberhard 1990: 342)—just the kind of intraspecific variation we see in language!

If this is the case, why do we hear nothing of "spider cognition"? Why might we not suppose that orb web–spinning spiders come equipped with some kind of "universal grammar" of web construction, complete with angles and relative proportions of height and breadth, perhaps incorporating knowledge of things like the tensile strength of viscous substances, basic properties of geometrical figures, effects of wind velocity, temperature, humidity, and the like, modifiable by experience and the example of conspecifics, but also allowing for individual modification, "idiolects" of web construction, one might say?

I think the answer is that bats and spiders perform their intricate operations via "instinct," while humans perform such feats as a result of "learning." The privileging of learning over instinct seems natural enough for a species in which learning plays as vital a role as it does in ours, but it is no less pernicious for that. Learning, along with cognition, culture, and consciousness, places the focus of inquiry on a single species, so that other species become of interest chiefly for the extent to which they foreshadow, albeit in crude and primitive forms, the achievements of humans. But the irony of this lies in the fact that language, the crowning glory of humans, and (according to Darwin; see Chapter 1) the likeliest cause of all our cognition, culture, consciousness, and hyperdeveloped learning powers, is largely instinctive.

Consider physical processes that we would unhesitatingly describe as instinctive, below consciousness, beyond cognition. Compare, for instance, the way we digest food with the way we produce sentences. Food enters our mouth, and that's the last thing we know about the process of digestion. Words leave our mouth, and that's the first thing we know about the process of sentence formation. We have some control over the content of our digestive processes (we can choose what we eat) but no awareness, unless they go badly wrong, and no conscious or volitional control over the processes themselves. We have some control over the content of our linguistic processes (we can choose what we mean) but no awareness and no control, save for the power to voluntarily start and stop them, over the processes themselves. We can no more decide "I'll put a relative clause first and an adverbial clause of purpose later" than we can decide "I'll digest the ice cream first and the hot dog later." There was a time when we could consume only milk; then we were weaned onto solid food and began gradually to ingest a wide variety of substances, but we don't remember anything about this process, least of all how it happened. There was a time when we could utter only funny noises; then the noises took on regular shapes and became words, and we began gradually to produce a wide range of sentences, but we don't remember anything about this process, least of all how it happened. So which of these processes is cognitive?

There's too much work in cognition that focuses our attention not on the processing of information per se but on information processing that is conscious and intentional. Hence while some attention is paid to animal processing, human information processing usually occupies center stage. Of the six long area articles with which Wilson and Keil (1999) commences, half ("Philosophy," "Computational Intelligence," "Linguistics and Language") deal with areas that by definition involve only humans and human inventions, while the rest ("Psychology," "Neurosciences," "Culture, Cognition and Evolution") concentrate mainly on humans. When nonhuman behaviors are under consideration, their relevance to human behaviors is of prime concern; typically "we are interested in the results [of experiments involving shock-induced inhibition of bar-pressing] only because we assume the rat is experiencing an emotional state related to human fear" (Ristau 1999: 133). But if cognitive science deals with information processing, it should be as concerned with how spiders know how to spin webs as it is with how humans know how to make sentences.

Humans are no more aware of what their brains are doing while they are speaking or listening to others than spiders are when they are spinning webs or bats when they are hunting insects. The fact that humans have large brains with unusual computational capacities does not mean that their cognition must somehow be implicated in the evolved infrastructure of language. If language is an evolutionary adaptation, the null hypothesis is that its infrastructure, as with spiders, bats, beavers, and all other species with complex behaviors, is purely instinctive, a blueprint for a set of algorithmic behavioral routines.

An "Instinct to Learn"?

But surely language differs from web spinning and bat echolocation in that the former contains a substantial learning component, while the latter doesn't? How we assemble words into sentences may be instinctive, but first we have to learn the words, and the sounds we use to make those words must be fine-tuned by experience. I don't know if isolation experiments have ever been carried out on bats or spiders, but my guess is that if a bat or a spider was raised without ever seeing another bat or spider, it would still be able to echolocate or spin a web as well as other species members. In contrast, children for whom some accidental circumstance has drastically reduced or eliminated linguistic input may never speak, or if they do may fall far short of a full adult language capacity (Curtiss 1977; Newton 2002).

Language might therefore be seen as more comparable to other kinds of nonhuman faculty, for instance birdsong. There are species of songbird whose songs are as preprogrammed as bat echolocation, but there are many others whose songs are at least partially learned. It has been known since the eighteenth century that cross-fostered members of some species can even acquire the song of the foster species (Barrington 1773), just as cross-fostered human infants may acquire a language totally different from that of their biological parents. More commonly, immature members of the species begin by producing first what is referred to as "subsong" and later something that has been termed "plastic song" (Marler and Peters 1982; Marler and Nelson 1992; Nottebohm 2005). Subsong has been compared to the babbling of human infants; plastic song represents the stages through which the young bird's production passes on the way to a full rendition of its species-specific song, comparable to the "two-word" and "telegraphic speech" stages in first-language acquisition (see Chapter 6). The ability to develop song in such species has been described as "an instinct to learn" (Marler 1991), a concept hailed by Fitch (2010) as potentially resolving the long-standing conflict between empiricists and nativists. Does this concept enable us to allow for the role of instinct in language yet still regard it as a cognitive function—to have our cake and eat it?

Language obviously depends in part on learning, otherwise we would all speak Humanese rather than English, Chinese, Swahili, or whatever. But the nativist-empiricist conflict won't be resolved by the kind of compromise that an instinct to learn represents. Language clearly falls into two parts. One part consists of the ways in which one language differs from another, and this part has to be learned. The other part consists of ways in which languages do not differ. It is this part of language that I am comparing to things like spiders' web spinning and bat echolocation. It does no good to mix up these two, as the "instinct to learn" formula mixes them. Humans don't have an "instinct to learn" the second part, the part that is like web spinning, because no learning is involved. They don't have an "instinct to learn" the first part, the part that results in language differences, because they don't need any specialized kind of instinct in order merely to learn what has to be learned if they are to become fully socialized beings in the particular society where fate has placed them.

Fitch (2011) gives a clear but, I think, mistaken view of how immature birds develop song which agrees with the popular picture of how immature humans develop language: "A young songbird, while still in the nest, eagerly listens to adults of its own species sing. Months later, having fledged, it begins singing itself, and shapes its own initial sonic gropings to the template provided by those stored memories." I strongly doubt whether young birds still in the nest are "eagerly listening" or doing anything much besides growing larger and making sure they get at least their fair share of incoming food. It seems much likelier that they are merely executing a biological plan and that later on, instead of adjusting "sonic gropings" to a template of "stored memories," they are adjusting that plan to fit the adult models they are simultaneously hearing.

Such a proceeding would seem to characterize both birdsong "learning" and language "acquisition." Nottebohm (2005) comes closer to getting both right when he suggests that "the variable mismatch between a model and the attempted imitation drives output modification, so that patterns that had not occurred before now first appear." All that I would question in transferring his proposal to human language is his use of the expression "attempted imitation." I would therefore rephrase the sentence as "The variable mismatch between the direct output of the biological program for language and data from a particular human language drives output modification."

To characterize the ontogenetic development of language in this way would align the development of humans with the development of other species. It would also, I believe, be nearer to the truth. Evolutionary processes apply in exactly the same way, across the board, to human and nonhuman species alike.

The "Component Features" Approach to Evolution

The special treatment of humans in evolutionary studies is not limited to the examples above. That special treatment is reinforced by the introduction of the "component features" approach. This approach owes much of its popularity to the fact that it is frequently characterized as an example of the comparative method, one of the most fundamental tools of evolutionary biology.

"Comparisons establish the generality of evolutionary phenomena.... From Darwin's time to the present, the comparative method has remained the most general technique for asking questions about common patterns of evolutionary change" (Harvey and Pagel 1991: 1–2). But in language there are few if any "generalities" or "common patterns"—language is a unique form of behavior. So what are researchers to do when they encounter such forms?

The one thing they have never done, except when dealing with language, is to take a particular unique trait or behavior of some species, break it down into what might appear to be its component parts, and then try to determine if those parts or analogs thereof can be found in other species, preferably close relatives, but if not, anywhere on the phylogenetic tree, with the goal of explaining how the unique trait or behavior evolved in the first place. Thus we do not hear of attempts to explain how orb web spinning developed in spiders by first breaking it down into component parts—thread production, passive trapping, and so on—and then looking at which other species, related or not, have managed to master these components. It's not impossible. One could examine thread production in silkworms and caddis fly larvae, or trapping by ant-lions, angler fish, the ant species *Allomerus decemarticulatus* (Dejean et al. 2005), and perhaps even carnivorous plants like sundews, and so on.

Or take bat echolocation—in its precise form, unique to bats. A number of other species use projected sounds for locating obstacles or prey, including toothed whales and dolphins, oilbirds, swiftlets, shrews, and tenrecs. Might not these shed light on how echolocation evolved in bats?

This possibility seems not to have even been considered. Work on bat evolution (e.g., Neuweiler 2003; Jones and Holderied 2007; Li et al. 2007) makes no mention of echolocation anywhere outside the immediate ancestors of modern bats, assuming without discussion that the trait evolved as a whole in response to circumstances peculiar to bats. But here, surely, one could make a much better case for a "component features" approach than could be made in the case of humans. After all, humans are one species, and languages, despite their numerous superficial differences, do not fall into distinct and clearly demarcated classes based on fundamentally different principles. Echolocation, in contrast, takes quite different forms in different bat lineages. Surely it would be quite natural to assume that the differences between forms of echolocation arose through different selections of component features, some perhaps found in tenrecs but not in oilbirds (or vice versa). We might reasonably expect to find a substantial literature on how each lineage had selected its form of echolocation from components found in the other species that used echolocation. But we find nothing of the sort.

However, one of the things we do find is a claimed role for the FoxP2 gene in echolocation, leading Li et al. (2007) to suggest bat echolocation as a precursor of human language! It is somehow seen as legitimate to ignore possible precursors for a particular behavior yet claim that behavior as the precursor of a second behavior, but only if the second behavior is something humans do. A clearer example of a double standard in biology would be hard to find.

Yet for Hauser et al. (2002), the assembling of "precursors" and "component features" is simply an example of biology's trump card, the comparative method. Their line of reasoning (1572) is worth following in some detail. They begin by claiming that the question of whether or not component features of the language faculty are unique to humans forms "an overarching concern" in language evolution studies. Since unique traits have often been attributed to humans on inadequate grounds, they state how essential it is to demonstrate, preferably by experiment, that other species do not possess them. If other species do possess them, they may be either homologous or analogous. For instance, many songbirds share with humans the feature of learned vocalizations, but "although the mechanisms underlying the acquisition of birdsong and human language are clearly analogs and not homologues, their core components share a deeply conserved neural and developmental foundation" (1572).

With regard to the "uniqueness issue," one needs to ask why that is treated as "an overarching concern" when human uniqueness is under consideration but not when dealing with uniqueness in other species. With regard to previously claimed "unique traits" that have been attributed to humans, their uniqueness has been disproven not by breaking down those traits into component features but by showing that other animals possess whole traits that are similar to the human trait, albeit perhaps less developed. For example, tool making was supposed to be unique to humans until it was found that chimpanzees make and use tools. Thus language remains unique to humans until someone can show that some other species possesses anything that linguists might regard as language. Nobody in the history of biology has treated any other species-specific trait as something that could be decomposed into a list of allegedly component features and exhaustively explained by combing nature for species that might exhibit similar features in dissimilar genomic, epigenetic, and ecological contexts.

The notion that language could usefully be divided into a set of discrete components seems to have originated with Hockett (1960; Hockett and Altmann 1968), who published a list of thirteen (later expanded to sixteen) "design features" of language, including such things as "semanticity" (linkage between sound and meaning) and "discreteness" (the fact that speech sounds are perceived as separate and distinct from one another, even though phonetically, especially in the case of vowels, they may be on a continuum). Hockett found almost all of these features in the behavior of one species or another. Hauser (1996: 48–49) had already written favorably about Hockett's features, although the features selected in Hauser et al.(2002) tend to be less descriptive and more functional (thereby underlining the fact that the choice of features into which a trait is decomposed is an arbitrary one, determined by the observer's own assumptions and predilections).

Thus there is every reason to be skeptical about the whole trait-decomposing approach. Any complex trait, whether it be web spinning by spiders, dam building by beavers, or sentence production by humans, has its own internal structure, its own external relationships with other traits, and its own application to a particular ecological challenge that impacts the species in question. Let us suppose, probably counter to fact, that it is indeed possible to factor out particular components of a given trait and find the same or similar components in the complex traits of other species. Take such a component, call it F, and say that it combines with components U, V, and W to yield a complex trait CT1 in species A. Suppose that F, or something resembling F, is found also in species B. But in species B, F combines with W, X, Y, and Z to yield complex trait CT2. Can F have developed in the same way? Is it not the case that what really evolved was CT1 or CT2 and that the developmental path followed by F, even the precise form taken by F differed markedly in the two cases, due to F's interaction with U, V, and W in the one case, and W, X, Y, and Z in the other? Things like "categorical perception" and "learned vocalization" are not valid, independent evolutionary categories. They are abstract categories whose various instantiations may have quite different developmental and evolutionary trajectories and serve quite different functions in different species.

However, Hauser at al. (2002: 1572), basing their position on their interpretation of "evo-devo," concluded that the various instantiations of F must "share a deeply conserved neural and developmental foundation."

Overestimating Evo-Devo

A few decades ago it was widely believed that individual genes played strictly specified roles—that the genes responsible for forming human arms were not the same as those responsible for the forelegs of mice, which again differed from those involved in the production of birds' wings. It followed that the more complex the organism, the larger the number of genes it should have. But as more and more species had their genomes sequenced (C. elegans Sequencing Consortium 1998; Adams et al. 2000; Arabidopsis Genome Initiative 2000; Rat Genome Sequencing Project Consortium 2004; and similar others) surprising facts emerged. Numbers of genes varied little between species (most animals, including humans, and many plants fall within a range of ~20,000 [+/− 10,000]), and the same genes turned out to be responsible for creating wings, forelegs, and arms.

Genome sequencing was a bonanza for the rapidly expanding field of evolutionary developmental biology, or evo-devo, which had its roots in nineteenth-century studies and was already well established when sequencing began (Goodman and Coughlin 2000). It helped confirm major claims in that field: that genes were far more pliable than had been thought, that gene expression could be influenced by the environment, that epigenetic factors played a significant role in development, and that developmental changes were powerful determinants of apparent evolutionary novelties.

The discovery of what came to be known as "deep homologies" (Shubin et al. 2009; Scotland 2010)—links between organisms that might be only distantly related, that would have been written off as analogies by previous generations, but where in fact identical genes were involved—led Hauser at al. (2002) to assume that the components of human language too could involve deep homologies. If this had been the case, their search for "precursors" or "component features" of language might have been legitimized. But most of those components were behaviors rather than structural forms, and evo-devo is principally about form. Carroll (2005: 4)—for many, including Hauser et al. (2002), the oft-quoted bible of evo-devo—states at the beginning of his introduction, "The initial inspiration for this story is the attraction we all share to animal *form*, but my aim is to expand that wonder and fascination to *how form is created*—that is, to our new understanding of the biological processes that generate pattern and diversity in animal *design*" (emphasis added). Indeed Hoekstra and Coyne (2007: 997), in an article critical of more extreme evo-devo claims, state that "advocates of evo-devo . . . make a sharp distinction between the evolution of the anatomy and the evolution of all other traits" and "the theory of gene regulation largely ignores adaptations affecting behavior."

Evo-devo is still young. The fact that it hasn't yet been able to produce an account of how behavior develops cannot be taken to mean that it won't at some future date come up with just such an account. Meanwhile there is good reason to be skeptical. Behavior is considerably further from direct genetic control than form is. This can be shown by simply considering the nature of behavior. Suppose we have a species X with a behavior Y. Capacity for a behavior inescapably depends on having the necessary form—a big enough brain, sufficiently developed organs of sense, limbs in the right places, whatever—and biological factors, genetic or epigenetic, mandate that form in all normal members of X. In other words, being a member of X mandates a capacity to perform Y. But capacity to perform Y does not mandate that Y will be performed.

Whether or not Y gets performed depends on one thing and one only: whether those members of the species that reliably and consistently perform Y make a contribution to the gene pool that is at the very least equal to the contribution of those that don't perform it. The environmental factors that make Y desirable or necessary are thus what ultimately control performance of Y, and not the form-determining functions of evo-devo or anything else.

In other words, natural selection is involved. Nobody but a pan-adaptationist straw man denies that there exist laws of form, scaling laws, and other nonselectional factors that are just as much determinants in evolution as is natural selection. But they operate in different ways and at different times. The factors described in evo-devo (and others) determine what kinds of organism can evolve in a world with the physical and chemical properties of ours. Natural selection applies after these organisms have evolved, determining whether their adaptations will become species-wide, linger on in a minority of the species, or disappear altogether.

Consequently to invoke evo-devo and deep homologies to justify a "component features" approach to language evolution is at best premature, at worst a blind alley. Complex traits, whether spider webs, beaver dams, or human language, evolve as wholes, not as an accumulation of ingredients, in species that require those traits to fully exploit whatever niche they have entered or created. And here is where evo-devo can really help us understand how this happens. What evo-devo most clearly reveals is the degree of plasticity in expression that enables the same genes to produce things as seemingly dissimilar as bat wings, bird wings, mouse forelegs, and human arms (Goodman and Coughlin 2000; Gilbert and Sarkar 2000; Carroll 2008). Linking evo-devo with niche construction theory (Laland et al. 2008) opens up new and revealing ways of approaching the evolution not just of language but of any new and complex forms of behavior.

Hauser et al. (2007: 108) do make at least one excellent point: "The goal [in dealing with possible analogies] isn't to mindlessly test every species under the sun, but rather, to think about the ways in which even distantly related species *might share common ecological or social problems*, thereby generating *common selective pressures* and ultimately, solutions given a set of constraints" (emphasis added). However, neither the article cited nor, to the best of my knowledge, any other work that takes a "component features" approach makes any attempt to carry out such a program. Indeed in the entire literature prior to Bickerton (2009), I know of only one author, Dunbar (1996), who has linked the evolution of language with any "ecological or social problem" peculiar to human ancestors. (I do not count frequent references to unspecified "social pressures," since these are never spelled out in sufficient detail to enable discussion.) For all other authors, language seems to have evolved in an ecological vacuum, a defect as surprising as it is deplorable.

Patching the Component Features

Hauser et al. (2007: 128–129) summarize their approach as a "patchwork quilt theory of language evolution," pieces in the quilt consisting of component features of language found in other species that were "woven together in humans by a species-specific syntactic innovation." Before we can accept such a theory, we have to accept that all the component features of language (except some "syntactic innovation," presumably recursion or Merge) somehow came together in either modern humans or some immediately antecedent hominid ancestor. But why would they? How could they?

For argument's sake, let us suppose (probably counter to fact) that this was somehow possible. By definition, all of those components must have been employed, in other species, for tasks other than language. Why would they not have simply continued, in humans, to perform the same tasks and nothing else? How and why would the mere addition of a single new component have woven the patches into a quilt? No explanations are offered.

If there was nothing like language until the "syntactic innovation" appeared, are we to assume that language emerged abruptly, in essentially its modern form? Such is the opinion of Chomsky, a coauthor of Hauser et al. (2002); "at some point, unbounded Merge must appear" (unbounded Merge being allegedly the product of recursion), and therefore "the assumption of earlier stages seems superfluous" (Chomsky 2010: 54). It is worth noting that another coauthor of Hauser et al. (2002) has a diametrically opposite view: Fitch (2010) devotes three out of fifteen chapters to discussion of various hypothetical models of protolanguage and favors a type of musical protolanguage originally argued for by Darwin (1871). Hauser's view on this issue is nowhere made explicit, but, like Chomsky's, it appears to necessarily rule out any direct involvement of natural selection in the evolution of language.

But no matter how profound the contribution of "deep homologies" or how many component features some early human had accumulated, nothing would have happened without natural selection. The various component features would have just gone on fulfilling their original tasks if there had not been some selective pressure conducive to some form of language. And language, in however embryonic a form, must somehow have increased the fitness of whoever had it, or none of the changes that supported it would have gone to fixation. Central to any research into how language evolved are questions concerning how and why language developed in one particular species rather than in others and why no other related species has made even the most modest move in the direction of language, despite having (presumably) a large subset if not all of the necessary component features. Indeed the higher the number of component features claimed to appear in other species, the more it becomes mysterious why humans and only humans developed language.

Not only do serious (perhaps the most serious) questions about how language evolved remain unanswered under the "patchwork quilt" theory, but that theory has some implications that any biologist would surely want to avoid. The notion that component parts of language assembled gradually over time until their mere collective presence triggered language suggests a progressive, teleological view of evolution more consistent with creationism or intelligent design than with naturalistic science. Indeed the patchwork quilt theory demands a Quilter, and if that Quilter wasn't natural selection, what was it? Yet Hauser et al.(2007) do not mention natural selection even once in the entire article, and mention selective pressures only twice, once in the passage cited above and once, at the end, as a kind of afterthought (128) in equally general terms. They seem to have taken it as self-evident that language was bound to be selected for, and this, alongside the implication that its component features were being gradually and steadily amassed over hundreds of millions of years, inevitably brings to mind the long-discredited *scala naturae* or divinely appointed Great Chain of Being.

Up the Ladder to Humans

I have been sharply critical of the "component features" approach, so it's only fair to mention one of its more positive aspects. It has served to draw attention away from what Pepperberg (2007) called the "primate-centric" approach, with its focus on "precursors" of language limited to species closely related to humans.

That a primate-centric approach should have predominated in language evolution studies for so long is unsurprising, given the coincidence of two powerful forces. The first was the belief that evolution must invariably proceed through innumerable small changes, and accordingly must be gradual and continuous. The second was the influence of what one might call "genetic triumphalism"—the notion, widespread in the decades immediately following the Crick-Watson unraveling of the genetic code and exemplified in Dawkins (1976), that genes were invariant and all-powerful agents of evolutionary change. Taken together, and given the genetic similarities between apes and humans, it seemed only logical to seek the immediate antecedents of language in primate ancestors. Since those ancestors were all long dead, the only course seemed to be the study of those primate species closest to humans, despite the vast behavioral differences between humans on the one hand and chimpanzees, bonobos, and gorillas on the other.

In fact similarities between closely related species are not always found even in the physical realm, let alone the realm of behavior. Consider three apparently very dissimilar species: the elephant, the hyrax, and the dugong. Based on their physical appearance, a natural assumption might have been that the elephant was most closely related to other large, thick-skinned mammals (hippopotamus or rhinoceros), that the hyrax was a small rodent, and that the dugong had somehow split off from earless seals. In actual fact, evidence from a number of sources reveals that the three are more closely related to one another than they are to any other species (Stanhope et al. 1998; Nishihara et al. 2005). Needless to say, to look for "precursors" of unique features, whether physiological (the elephant's trunk) or behavioral (its low-frequency communication), in hyraxes or dugongs would be quite futile. But the greater degree of physical similarity between apes and humans seems to have led researchers to assume that comparable behavioral similarities must also exist.

In fact research along the lines recommended by Hauser et al. (2002, 2007) has indicated that even if we take a component features approach, there are numerous features for which other, more remotely related species seem far better adapted than the great apes. "Functional reference" (expressed in calls used to warn conspecifics of approaching predators), seen by some as a precursor of words, is found in vervet monkeys (Cheney and Seyfarth 1988) and even chickens (Evans et al. 1993) but is absent from the communication of great apes. Ape vocal learning, as well as capacity to reproduce speech sounds (Hayes and Hayes 1952), is inferior to that of parrots (Pepperberg 1991), dolphins (Richards et al. 1984), and even harbor seals (Ralls et al. 1985). Yet such differences are not merely ignored but sometimes even denied, despite empirical evidence to the contrary. In discussing the absence of functional reference from ape communication, Mithen (2005: 113) states that "it is in this African ape repertoire, and not in those of monkeys or gibbons, that the roots of human language and music *must be found*. The *apparently* greater complexity of monkey calls *must be an illusion,* one that simply reflects our limited understanding of ape calls" (emphasis added).

Mithen's (2005) remarks are based on a widespread if seldom explicitly stated assumption: that animal communication has evolved by the kind of ratchet-like process that produced eyes or wings—a process developing from simple beginnings but preserving each improvement until the end product is a complex organ exquisitely adapted for whatever it does. But as already noted, physiology and behavior evolve in quite different ways. The signal repertoires of some species—it seems inappropriate to call them "systems" since they have no internal structure—contain more units than those of others, but there is no consistent correlation between species complexity and number of signals (Wilson 1972). Unlike language, no signal repertoire has any kind of internal structure, no equivalent of even the simplest grammar. Indeed there is no sign of "progress" in any direction, let alone progress in the direction of human language. But again humans (and anything connected with humans) are treated as special. Only when a human connection forms the object of inquiry is a preexisting signal repertoire supposed to become

58   The "Specialness" of Humans

capable not just of evolving but of evolving into a complex system unlike anything else in nature.

A moment's thought should show why physiological and behavioral evolution necessarily differ from one another: it is for the same reason that evo-devo has been able to account for physiological but not behavioral adaptations. Physiology is under much tighter genetic control than behavior (although even physiology may be freer than commonly supposed; see recent work such as Gilbert and Epel 2009 or Piersma and van Gils 2010). "Behavior is the result of a complex and ill-understood set of computations performed by nervous systems and it seems essential to decompose the problem into two: one concerned with the question of the genetic specification of nervous systems and the other with the way nervous systems work to produce behavior" (Brenner 1974: 72). The problem is even more complex than Brenner could have known. He wrote before either evo-devo or niche construction theory had developed, in other words, before genetic plasticity was understood and before it was realized that behavior too has degrees of freedom. Behaviors are primarily determined not by factors intrinsic to the organism but by the exigencies of the niche that a species occupies.

In other words, the behaviors of species are rather precisely tailored to fit, not the environment in general but the *Umwelt*, those aspects of their environment with which given species directly interact (Uexküll 1910, 1926). Given the degree to which, as evo-devo has shown, environmental factors can influence gene expression, one might expect that the components of complex novel traits would develop in place, in the species that required those traits, in the course of the development of those traits. One would certainly not expect that such components would have to be handed on, like batons, ready-made or at least capable of being perfected, from one species to another. But that is exactly what is presupposed by both the component features and the primate-centric approaches.

Critics of a focus on primates and precursors are often accused of "typically tak[ing] the Cartesian position that language is special, in the sense that all of its attributes are unique in humans," as well as of claiming that animal signals are "simply a read-out of emotional states," produced in a "reflexive and involuntary" manner (Evans et al. 1993: 315). Neither of these accusations applies here. Any claim that *"all* of [language's] attributes are unique in humans" would be absurd, if anyone actually made it. All that the antiprecursor position claims is that the existence in other species of phenomena similar to some found in language tells us nothing of significance about how language evolved in humans. Nor has the uniqueness (or otherwise) of language any connection with the belief, known to be false since at least Cheney and Seyfarth (1990), that animal signals are involuntary expressions of emotion. I reject the "primates and precursors" approach not because it threatens human specialness but for a precisely opposite reason: because it covertly reinstates human specialness. Just like the component features approach, it treats humans in ways quite different from those used when nonhumans are objects of study.

Treating nonhuman behaviors as precursors of human behaviors has been well described by Rendell et al. (2009: 238), who in reviewing comparisons between language and animal communication state, "Although the loosely defined linguistic and informational constructs make convenient explanatory shorthand, they are problematic when elevated beyond metaphor and pressed into service as substantive explanation for the broad sweep of animal-signaling phenomena." In reality, animal communication is designed primarily to manipulate, not to inform (Krebs and Dawkins 1984). Rendell et al.'s criticism, far from being "a declaration of evolutionary discontinuity," is quite the reverse. It is motivated by the fact that "characteristics of signaling in an array of species are routinely tested for possible language-like properties, thereby turning the normal evolutionary approach on its head. The equivalent for locomotion would be to take the mechanisms, functions, and energetics of human bipedality as a model for understanding the quadrupedal condition from which it evolved" (Owren at al. 2010: 762). Owren et al. conclude that "if the mechanisms . . . are fundamentally different, the parallels that are so often drawn exist primarily in a metaphorical rather than a real-world domain. We suggest that they are therefore more a distraction than a boon to serious scientific inquiry" (763).

Consequently the search for features of animal communication that prefigure components of language is "both teleological and circular" (Rendell et al. 2009: 238), a claim that seems fully justified. Consider exactly what is entailed by saying that the words and syntactic structures of language have "precursors" in animal communication. It has been claimed, for instance, that the alarm calls of vervet monkeys (Cheney and Seyfarth 1990) are word-like. More recently there have been claims of syntax-like (or at least combinatorial) utterances by putty-nosed monkeys (Arnold and Zuberbuhler 2006). Researchers who make such claims or who endorse them (e.g., Hurford 2007a; Kenneally 2007) are looking at animal signals not as phenomena in their own right but as indicators of the extent to which other species anticipate aspects of human language. It is as if language represented some kind of adaptive peak that other animals were struggling to climb (but were unsuccessful to the extent that they failed to attain it). Again this inevitably recalls the *scala naturae* or Great Chain of Being, the divinely sanctioned pyramidal structure that formed the prevailing view of nature before Darwin changed it forever. The irony is that those who seek "precursors of language" sincerely believe they are reducing the distance between humans and other species.

Primates and Pressures

We have reviewed two ways of approaching human evolution that treat humans and nonhumans differently. Unfortunately there is at least one more. What is bizarre about this one is that it makes humans special by effectively denying them any role in their own evolution.

As noted earlier, most books on the evolution of language make no attempt to integrate it into the overall evolution of humans, despite the fact that the evolution of language forms an essential and (I argue in these pages) the single most important part of that process. Either hominid ancestors are ignored altogether or human evolution and language evolution are treated as entirely separate topics. As an example of the first approach, Hurford (2007a) includes only a page about sexual dimorphism in australopithecines and a couple of tangential remarks about *Homo habilis* and *Homo erectus* (both in citations from other authors) in a work of nearly four hundred pages. As an example of the second, Fitch (2010) devotes three chapters (nearly a hundred pages) to an account of human ancestry (starting from single-celled organisms), nearly fifty pages of which concern our most recent ancestors. But this chapter (significantly entitled "Hominid Paleontology and Archeology"—not hominid behavior!) is almost entirely limited to the usual "bones and stones" account found in paleontology textbooks, and Fitch's own version of the earliest stages of language is not presented until 170 pages later. There are no cross-references with his "hominid" chapter and no serious attempts to ground his hypothesis of a "musical protolanguage" in specific behaviors of any human ancestor.

But perhaps the most telling way to demonstrate the extent to which human ancestors are eclipsed by other primate species in the language-evolution literature is to present in tabular form the total number of page references to extinct ancestral species found in a representative sample of such books as compared to the number of page references to living primates, as shown in the index of each book. As Table 3.1 shows, works on language evolution refer to living primate relatives that did not develop language on average more than five times as frequently as they refer to species one of which must have actually developed language.

How are we to explain this surprising disparity? The obvious response is to say, "Well, we know relatively little about human ancestors, who aren't available for study, but a lot more about living primates, who are, and because of the high degree of genetic closeness between humans and primates, we can legitimately use primates as a model for prehumans." In other words, it's a classic case of looking for your car keys where the streetlights are.

Table 3.1. Number of page references to hominids versus other primates in language-evolution studies

	Hominid	Other primate	Ratio H:OP
Aitchison (1996)	11	31	1:3
Deacon (1997)	67	117	1:1.75
Burling (2005)	17	74	1:4.3
Kenneally (2007)	40	217	1:5.5
Hurford (2007a)	8	180	1:22.5
Total	143	619	1:5.3

Using modern ape behaviors as a baseline for subsequent hominid development commits us to viewing whatever developments led to language as being no more than enhancements of traits already possessed by apes. By many, this is seen as both natural and desirable. In the preface to a recent volume devoted to considering the relationship of humans to other apes (Kappeler and Silk 2010: v), the editors state—as fact, not argument—that "comparative studies of our closest biological relatives, the nonhuman primates, provide the logical foundation for identifying human universals as well as evidence for evolutionary continuity in our social behavior." However, this currently extremely popular view is pernicious in several respects. First, it trashes more than five million years of evolutionary development in the line that led directly to *Homo sapiens*, assuming without argument that nothing significant can have happened in that interval. Second, although aimed at reducing the "specialness" of humans and emphasizing human continuity with other species, it treats apes not as well-adapted animals in their own right but rather as incomplete humans—organisms, to use Jane Goodall's (1971) telling phrase, "in the shadow of Man." Third, it drastically reduces our capacity to answer a key question in language evolution: Why did humans and no other species develop language?

The more precursors and component features of language we uncover in other species, the harder it becomes to answer this question. As noted earlier, regardless of which scenario one chooses, natural selection has to apply at some stage. Even the Promethean mutation of Chomsky (2010) would never have reached fixation if Prometheus's descendants hadn't outbred the competition. But the choice of modern apes as models for ancient humans limits candidates for the crucial selective pressure(s) to those that would also have affected other apes.

The number and variety of pressures that have been hypothesized (Johansson 2005 lists over a dozen, from tool use and hunting to social intelligence and female choice) should be enough in itself to excite suspicion. None stands out; none has a predominance of evidence over others. None is more than a Hail Mary hypothesis, a desperate attempt at forming *some* proposal in light of the supposed lack of relevant evidence and the extreme difficulty of the problem given this supposed fact. But all these theories fail a simple thought experiment. Bear in mind that (proto)linguistic behavior had to pay off from the earliest utterances or its genetic infrastructure would never have reached fixation. So consider, given your favorite selective pressure, what those first utterances *could have been* and whether they could have been *useful enough* to be retained and added to.

Note that although this sounds like asking "What were the first words?," it is not the same. The question is aimed simply at finding out whether, given a certain selective pressure, it would have been possible to produce utterances that would have been in any way relevant to that pressure. Take the "grooming and gossip" hypothesis of Dunbar (1996). This does at least satisfy the uniqueness-to-humans condition in that it claims that an increase in group size among human ancestors made the grooming of significant others too time-consuming and therefore caused it to be replaced by the sharing of titillating bits of information about other group members. Such information would require not only signs for particular group members (no big deal: dolphins can do it; see Janik et al. 2006) but also signs for interactions between group members. Utterances like "X fought/had sex with/stole food from Y" require not only verbs—much more slippery meaning-wise than nouns and therefore seldom found among the first fifty words acquired by children—but enough syntax to distinguish "X did Y to Z" from "Z did Y to X." None of these would have been available until language had undergone a considerable degree of development. In other words, the earliest stages of language could not have expressed any gossip of the slightest interest, and since some other motive would have had to drive those stages, gossip as a grooming substitute cannot have constituted the selective pressure underlying the origin of language.

Viability of first utterances is far from the only thought experiment that can be used to evaluate selective-pressure hypotheses. Four more can be found in Szamado and Szathmary (2006), where eleven hypotheses are examined and all found to fail at least one of those tests. However, there is a much stronger argument for rejecting pressures that would have affected other species. Where the human species is not involved, it is assumed (even by some who give special treatment to humans) that novel traits arise as responses to novel problems that are faced by particular species. For instance, Hauser (1996: 154) asks, "What *special problems* do bats confront in their environment that might have selected for echolocation?" (emphasis added). He does not discuss reasons for asking this question. To the contrary, he appears to regard it as self-evident that this is what one does when dealing with any novel adaptation. Indeed I have already cited a case (Hauser et al. 2007; 108) where he explicitly states the "special problem" criterion as a general research strategy. And yet when language rather than echolocation is the topic, as in Hauser et al. (2002), no mention is made of any "special problem" that might have selected for it.

The force of this objection will seem even stronger when we take into account niche construction theory (see below). It will then become clearer *why* novel adaptations can arise only through coping with "special problems." Researchers who focus on pressures affecting other apes may believe they are honoring evolutionary principles of gradualness and continuity, but in so doing they are dishonoring some of the most central evolutionary facts. One is that the genetic variability on which evolution works—without which there could be no evolution—exists mainly because organisms live in a wide variety of environments, rely on a wide variety of nourishment, and obtain that nourishment in a variety of different ways. Another is that the nature of environments, available nourishment, and means of procurement are subject to constant changes and that the responses of organisms to these changes are the motor of evolution, without which life could hardly have developed beyond bacteria. And though we do not know as much as we would wish about human ancestors, what we do know is enough to show that their environments and ways of exploiting those environments diverged very far from anything found among other primates.

Niche Construction Theory

Evo-devo is not the only new development in twenty-first-century biology. Niche construction theory (Odling-Smee et al. 2003; Laland and Sterelny 2006; Laland et al. 2008) is equally significant.

Proponents of niche construction theory argue that classic neo-Darwinism (Mayr 1963; Williams 1966: Ayala and Dobzhansky 1974) unduly restricts the role played in evolution by the evolving organisms themselves. From pond scum on up, living organisms actively cause changes in their environment by adding to, subtracting from, or modifying various aspects of that environment.

Everyone (including even creationists) accepts that if the environment changes, animals will adapt to those changes. According to some noted evolutionists, that is all there is: "Adaptation is always asymmetrical; organisms adapt to their environment, never vice versa" (Williams 1992: 484). And in many cases organisms do indeed adapt to the environment in which they happen to find themselves. But often that environment is radically altered, for example by abrupt and severe climate change or the emergence of new predators. Organisms then have three options: they can go extinct; they can adapt to the new conditions; or they can move into a new niche. (Note that though the last option often involves a move into new territory, it doesn't need to; members of the species concerned can remain in the same territory but change the way they exploit that territory by changing what they eat or their method of obtaining it.)

The third alternative has been chosen many times in the course of evolution. Aquatic animals have come on land. Terrestrial animals have returned to the water. Surface-dwelling species have climbed into the trees or gone underground. The first to move were, by definition, unadapted to their new terrain. However, the very fact of their having moved changes the evolutionary pressures that bear on them. It is at this point that the concerns of niche construction theory begin to overlap with those of evo-devo (Laland et al. 2008), since the newly understood plasticity of genetic factors is precisely what enables an organism to adjust rapidly to new conditions. A feedback effect is thus created whereby organism and niche mutually and simultaneously act upon one another, the organism transforming the niche, and the niche in turn transforming the organism until an almost perfect (or at least "good enough") fit is achieved between organism and niche—precisely the kind of fit that leads creationists to hypothesize a Creator.

For example, the adaptation of a previously terrestrial species to a mostly aquatic life that involved damming streams didn't require the beaver to assemble the teeth, eyes, claws, fur, tail, and so on necessary for that life before venturing into the water. To the contrary, it would appear that the beaver's eyes are both unique and exceedingly well adapted to the beaver's (constructed) niche. In other words, novel adaptations typically result not from an accumulation of preexisting components but from the evolution in place of a suite of components which, while they might find analogs among other species, were the product of interaction between a particular organism and a particular niche.

Note that niche construction theory serves, among other things, to sharply reduce the apparent differences between humans and nonhumans. Before further examining the links between niche construction and speciation, it is worth taking a brief look at one of these, something we refer to as "culture." The word has been used with a variety of meanings, some of which ("refined and sophisticated works of art and intellect," "patterns of behavior based on symbolic capacity") effectively limit use of the term to the human species. Probably most in the behavioral sciences would agree with something along the lines of "behaviors that vary between groups within the same species and have to be socially learned." Such definitions have licensed accounts of "culture" in a variety of species (Bonner 1983; Avital and Jablonka 2000), but especially among apes (Wrangham 1995; Byrne 2007). However, the disparity between the number and diversity of cultural practices among humans and their rarity and paucity in other species makes the usage look more like a desperate attempt at political correctness than a genuine effort to establish continuities of evolutionary process.

However, if instead of calling it "culture" we regard the whole range of variable human behaviors as simply an example of niche construction, we place humans on a continuum that links them with many other species, including some as phylogenetically remote as termites (Abe et al. 2000). Ants and termites have constructed elaborate niches; humans have developed an even more elaborate niche and by assigning it the name "culture" have set up yet another mechanism for making themselves look special.

Some may object that the reduction of culture to a product of niche construction is illegitimate because human culture involves learned behaviors while niche construction in other species involves mostly unlearned behaviors. But there are a number of responses. First, as noted earlier, the privileging of learning over instinct—almost an assumption of its moral superiority, one is sometimes tempted to think—has no rational basis. Second, niche construction can't involve genetically based, instinctive behaviors. The very novelty of a created niche (i.e., its novelty for the species constructing it) guarantees that its creators don't yet have any special adaptation for it. New niches can be developed only through behaviors for which there is no prior instinct, behaviors that have to be invented and passed on by learning before they can trigger the adaptive pressures that will eventually ensure them a place in the genome. Third, recall that virtually the entirety of human culture depends on the possession of a single faculty, the faculty of language, the workings of which lie as far from conscious awareness and volitional control as any of the "instinctive" behaviors of other species. In other words, all of our elaborate "learned" "culture" rests on a foundation as instinctive as that which underlies dam construction by beavers or fungus farming by ants.

Given such facts, it is unbecoming for humans to boast, like the fly on the chariot wheel, "Look at the dust I raise!" Indeed without niche construction, we could never have come into existence as the species that we are.

Niche Construction and Hominid Speciation

"One of the ironies of the history of biology is that Darwin did not really explain the origin of new species in *The Origin of Species*, because he didn't know how to define a species. *The Origin* was in fact concerned mostly with how a single species might change in time, not how one species might proliferate into many" (Futuyma 1983: 152). Under the genetic blending that Darwin assumed as the result of assortative mating, it is indeed hard to see how speciation could take place. Mendel's research radically changed that picture, and the theory of speciation is now a well-studied field.

Speciation can be allopatric (involving the physical separation of two groups from the same species) or sympatric (where both groups continue to inhabit the same territory). Here we shall be concerned only with sympatric speciation (Maynard Smith 1966; Dieckmann and Doebeli 1999; Higashi et al. 1999; Gavrilets and Waxman 2002). Unsurprisingly most speciation is allopatric. However, there is good reason to believe that hominid speciation was primarily if not exclusively sympatric (Summers and Neville 1978).

68   The "Specialness" of Humans

If one looks at a map of archaeological sites in East Africa, particularly in the Rift Valley, sites associated with different species in the human lineage can be found clustered together, suggesting that the same areas were used by several species over several million years. Moreover recent findings indicate that *Homo habilis* and *Homo erectus* (species previously thought to be ancestral to one another) overlapped and coexisted for perhaps as long as half a million years (Spoor et al. 2007). There can be a number of different causes for sympatric speciation, but some of the most powerful come from differences in resource exploitation; either different resources are used, or the same resource is obtained by different means and/or is differently used (see sources cited earlier). Changes in resource exploitation occur frequently in the context of niche construction, which may cause an organism either to enter a preexisting niche (Kawata 2002) or actively seek to create a new niche (Beltman et al. 2004). In the second case, "The colonization of a new niche is a first step towards speciation . . . as a result of selection for adaptation to the new niche" (Beltman et al. 2004: 36).

However, once this first step is taken, progress usually continues through subsequent stages of speciation to a point where hybridization becomes vanishingly rare or impossible. By the time that happens, species and niche have usually achieved a good fit, and subsequent changes are rare and usually modest. Laland et al. (1999) were perhaps the first to point out the intimate connection between niche construction and punctuated equilibrium (Eldredge and Gould 1972). Eldredge and Gould pointed out that typically in the paleontological record, new species appear relatively suddenly and then persist with little or no change over relatively long periods. This proposal was attacked by gradualists such as Dawkins (1986), who proposed instead something he called "variable speedism." But neither side in the controversy produced any explanation of why either "punctuated equilibrium" or "variable speedism" should exist.

The facts themselves support the Eldredge and Gould (1972) claim rather than that of Dawkins (1986), and niche construction explains why this should be so. A species needs to adapt rapidly if it enters a new niche. But once that niche is developed, there is little need for further change—unless there is some drastic change in the environment, in which case another speciation event becomes possible. Chapter 4 takes up this issue in a more detailed examination of hominid speciation. Here we need only note that evolutionary novelties are most likely to emerge in the early stages of speciation, and they are unlikely to occur except during the construction of some novel niche. Reasons for this become more apparent in the next section.

Some Guidelines for the Study of Language Evolution

I have made some negative comments on the way Hauser and his colleagues have approached the issues. It is therefore only fair to say that their work also contains two important pointers (not, unfortunately, followed by them) that, if taken together, will help to steer us in the right direction.

The first, from Hauser et al. (2002: 1572), advocates "the extension of the comparative method to all vertebrates and *perhaps beyond*" (emphasis added). The second, already cited and essential for constraining the first pointer, directs attention to "ways in which even distantly related species *might share common ecological or social problems,* thereby generating *common selective pressures* and ultimately, solutions given a set of constraints" (Hauser at al. 2007: 108, emphasis added). Note that it is only searches for "precursors" or "component features" that I have criticized as an abuse of the comparative method. Searching for "common problems" and "common selective pressures" is another matter entirely, since it focuses on similarities between niches developed by different species rather than on traits developed by other species so as to better adapt to their own particular niches. In addition it focuses on evolutionary convergence (Conway Morris 2003), the fact that there seems to be a limited number of solutions for any evolutionary problem. Good solutions to any given problem will therefore tend to be repeated, regardless of large species differences. These differences will necessarily cause variance in the precise means by which a particular solution is achieved, but that is only to be expected. Phylogenetic distance is no longer the embarrassment that Mithen (2005) and others feared.

Following these two pointers removes any need to assume that other primates' behaviors must necessarily be taken into consideration. If other primates had faced "ecological or social problems" similar to those faced by human ancestors, they would probably have attempted similar solutions. If the shared problem had anything to do with communication, their responses and ours should both have entailed some change in the means of communication. But there is not the slightest sign that the communication patterns of apes have changed in any way since ape and human lineages parted company—certainly not any that would point in the direction of language. This fact clearly indicates that we should abandon apes as models of human evolution. Instead we should look at the history of more immediate ancestors for a problem unique to those ancestors that had to be faced at a date later than that of the last common ancestor of humans and chimpanzees.

Next we should face the consequences of assuming that, as suggested in Chapter 1, we may have to deal not with a single process ("the evolution of language") but rather with two or more distinct processes of which language was merely the end product. We should look for whatever process took human communication away from the stimulus-bound signals of other animals to the threshold of the stimulus-free world that language would open up. This might have been easier when the issue was most frequently described as "the *origin* of language." Rephrasing this as "the *evolution* of language" seems somehow to have taken the focus away from the very earliest stages. Hurford (2007a: 333), for example, writes, "As soon as the breakthrough was made for animals to communicate their thoughts relatively freely to others, a cascade of other innovations were [sic] selected." We are not told what that breakthrough was, nor when, how, or why it happened, nor what innovations were contained in the cascade.

It could surely be argued that the failure to focus on origins, as much as any intrinsic difficulty in the topic, is what has made the question of how language evolved "the hardest problem in science" (Christiansen and Kirby 2003: 1). Even if language evolution had been a single process, any process has to have a beginning, and once you know how a process began, it is relatively easy to figure out its subsequent stages. The oft-repeated mantra that the origin of language cannot have had a single cause but must have involved the combination of many different factors goes a long way toward explaining why the problem has been made to seem so intractable. Nobody denies that a number of things had to be present before any species could aspire to language or that many diverse factors were involved *at some stage in the emergence of language.*

Those statements, though true, are profoundly unhelpful. One is left with a laundry list of possible factors but no metric for ranking their relative importance nor any means of determining the order in which they became operative. No wonder the problem looked hard. But however many factors there were, they must have applied in some particular sequence that only a carefully crafted scenario of the earliest stages of the move towards language can reveal.

Accordingly, inquiry should begin by looking at any circumstances that might have driven human ancestors to go beyond the limitations imposed by most animal communication. We should beware of any theory claiming that some internal factor such as a genetic mutation could be a sufficient condition for the emergence of any evolutionary novelty.

Significant evolutionary novelties do not result from spontaneous changes in the organism. Instead, as niche construction theory shows, they come about when changes in behavior develop as adaptations to particular niches. Take one rather obvious example: the origin of avian flight. There have been at least four theories (Zhou 2004). The cursorial theory proposes that bird ancestors ran faster and faster and made longer and longer leaps. The arboreal theory proposes that they jumped from bough to bough, eventually gliding, like modern flying squirrels. More sophisticated modern theories propose either that they were predators who pounced on terrestrial prey from branches (Gartner et al. 1999) or that they made ground nests but retreated to nearby branches at night, diving to the rescue if eggs or offspring were in danger (Kavanau 2007). Diverse as these theories are, they share one factor: all agree that some behavior came first and that the genetic adaptation of form followed it. No one ever suggested that flight began when some random mutation started wings growing.

We can now specify rather precisely the only conditions under which language could have been evolvable. Some species younger than our common ancestor with the apes and directly ancestral to modern humans must have confronted some very specific problem. That problem must have had two clear-cut and indisputable characteristics. It must have been a problem that no other species of remotely similar mental capacity had to face. And, whether it was "social" or "ecological," it must have been one whose solution involved some kind of change or addition

72   The "Specialness" of Humans

to the communicative behavior of prehumans. Of course, identifying such a problem is far from completing the story. It has to be shown that solving the problem would have far-reaching effects on the species that solved it—effects that could hardly have been foreseen, and that even today evolutionary linguistics seems hardly aware of.

CHAPTER 4

From Animal Communication to Protolanguage

Maynard Smith and Szathmary (1995) saw language as the eighth and latest of a series of major transitions in evolution. Two of the others are the emergence of multicellular organisms and the origin of life itself, so language moves in some pretty heavy company. And clearly events of such magnitude cannot occur like the flip of a switch. To the contrary, how life began breaks down into several different questions (Dyson 1999): how reproduction began, how metabolism began, and how replication began. (Note that, as Dyson explains, reproduction and replication are two different things.) And whether multicellular organisms had symbiotic, colonial, or syncytial (single cells with multiple nuclei) origins, their emergence was a process involving a series of intermediate steps and taking place over a substantial period (Hedges et al. 2004).

Thus if, as most assume, the transition from an alingual state to modern human language formed a single unitary process, that transition, whether gradual or catastrophic, represented a development unknown in any other evolutionary context. Only an account of language evolution as a series of dynamic processes can avoid the trap of human specialness and provide a coherent explanation of how language evolved. Just as in Dyson's account of how life began, we have to answer three questions—in this case, how the bonds of animal communication were broken, how the brain imposed structure on the output of this first process, and how, despite having a single basic structure, the final result was not a single language but several thousand languages. This chapter seeks to answer the first question

In dealing with any process, it helps to start with some kind of baseline. That baseline should have two components. Obviously it's important to know where to start communicatively. But it's equally important to know what there was to communicate—how rich (or how poor) was the cognitive state of our last alingual ancestors—because the relative quality of prelinguistic cognition must surely influence the subsequent course of language development. If cognition was initially very rich, comparable in every way to that of modern humans, a quite short trajectory for that course would have been possible. If prehumans were "cognitively, pretty much as you are" (Hurford 2007a: 164), with concepts similar in all respects to those of humans, then (leaving aside for the moment syntax and phonology) all that had to be done was find labels for those concepts. If, on the other hand, cognition was relatively poor—if our ancestors could mentally represent only what they could convey communicatively, what Chomsky (2010: 57) called "mind-independent objects or events in the external world"—a lot more work would have to be done. Language as we know it depends crucially on prior possession of symbols that pick out not particular individuals or occurrences but general categories of object and event—not "that black dog" or "running to the store last Friday," but simply "dog" or "running."

With regard to a communicative baseline, we can assume that communication in the last alingual ancestors of modern humans differed little from communication among living apes (Mitani 1996; Pollick and de Waal 2007). It would thus have consisted in at most a few dozen innate signals expressing either desires or intentions of the signaler or attempts to manipulate the behavior of conspecifics. All signals would have been inextricably bound to particular types of situation and particular contexts of utterance. (Use of signals to deceive does not count as an exception; for deception to work, signal receivers have to believe that context and situation are appropriate.) No signal would have been able to convey objective information.

Some may object that functional reference (discussed in Chapter 3) is frequent in the communication of organisms from vervet monkeys to chickens, so that objective information about the presence of predators is thereby conveyed. But this belief rests on the assumption that warning signals are in some sense "names" for the predators warned against. Typically, however, animal signals do not express anything we could paraphrase with single words. If we try to translate them into Humanese, we find the nearest equivalents are whole (normally imperative) clauses: "Mate with me," "Keep off my territory." The vervet interpretation of the "eagle" warning is probably better represented by some expression along the lines of "Look out, danger from the air!" or "Quick, run into the bushes and hide!" In any case, like all signals that do not merely express the signaler's current feelings, the intended function of the signal will be to manipulate behavior rather than inform. Of course there is nothing to stop animals from inferring information—in this case, the presence of an eagle—just as we do when we say things like "Those clouds mean rain by the evening." But we are not reduced to mere inference by the kind of utterances typical of human language.

The communicative baseline is not difficult to establish. The cognitive baseline is another matter.

The Paradox of Cognition

A recurring concern in post-Watergate politics has been "What did X know, and when did X know it?" That concern is at least equally pressing when we consider cognition in human ancestors. Here the problem is compounded by the fact that evidence of seemingly equal plausibility can be assembled on either side of the debate. One large body of evidence shows that nonhuman organisms must have a large store of rich and complex concepts. Another large body of evidence shows that there must be some vast cognitive gulf between humans and all nonhuman species, expressed in terms of things that the former can do that the latter cannot. In order to resolve this seemingly paradoxical situation, let us weigh the evidence on both sides of the debate. In what follows I deal only with concepts of concrete, nameable entities. More abstract relational concepts, such as same-different, will not be considered, even though some of these suggest impressive capacities in many animals. All that is at issue here is a capacity to hold concepts of a kind that could, potentially at least, be represented by words.

Evidence for Advanced Cognition

To many, one of the strongest arguments in favor of advanced cognition among prehumans has been evolution itself. They see "descent by modification," Darwin's preferred term for evolution, as virtually entailing that each modification be extremely slight. There is therefore not sufficient time for a vast increment in cognition to arise between the last common ancestor of humans and chimpanzees and the appearance of modern humans. Consequently at the dawn of language there can have been relatively little difference between the cognition of human ancestors and their closest relatives.

Darwin (1871: 57) himself supposed that "the mental powers of some early progenitor of man must have been more highly developed than in any existing ape, before even the most imperfect form of speech could have come into use." The view that increased intelligence is a prerequisite for rather than a consequence of language is widely shared by modern writers. Burling (2005: 72) states flatly, "Even before [human ancestors] began to speak, they must have dealt with the world by means of a rich set of concepts. Names cannot be given to things and events unless the learner already has the ability to form concepts for those things and events." Statements like this seem to discount the possibility that language and cognition developed in a coevolutionary spiral.

Turning to empirical evidence, it has been known since the 1960s that even pigeons can form concepts of a wide variety of natural and artificial objects and of other organisms, not merely ones they may already be familiar with (e.g., humans) but ones of which they almost certainly had no prior experience (e.g., fish; Herrnstein 1985; Pearce 1989; Watanabe 1993). These were, of course, experimental results. But the formation of concepts also takes place in the wild. Scrub jays (Clayton and Dickinson 1998; Emery and Clayton 2001; Raby et al. 2007) show that concepts of other species may persist in their minds without any stimulus from the conceived object. Scrub jays provisioning for winter establish thousands of food caches in different spots and then recover the contents months later, eating the most perishable first. They even keep track of whether they are watched by other birds during caching and rebury the food if this occurs (Dally et al. 2006). This suggests an expectation that others will steal the cache if not prevented from doing so—quite a sophisticated piece of mind reading for a bird.

If jays and pigeons can form and maintain concepts, we would expect apes to do at least as well. Experiments by Kohler (1927) showed that apes were not restricted in problem solving to objects they could currently see but would retrieve boxes or other objects from different sites in order to obtain fruit placed out of their reach. This could, of course, be written off as mere recall of specific objects and their locations rather than possession of abstract categories of object or location; possession of full-fledged human-like concepts is much more difficult to demonstrate. It is not so easy to write off the performances of "language"-trained apes. One of the most striking of these was the training of the two chimpanzees Sherman and Austin to use lexigrams not merely for concrete classes of objects (e.g., different kinds of tool or food) but also for the abstract, superordinate classes "food" and "tool." This suggests ability to acquire at least one aspect of symbolism: relations between and ranking of concepts.

Some authors (e.g., Chater and Hayes 1994) have concluded somewhat pessimistically that it is premature to claim human-like concepts for nonhumans since human definitions of concept are inextricably linked with language and because we do not know how animal concepts are neurally instantiated. This view, though understandable, seems unduly restrictive. Provided an animal has some representation of a class of entities X that can be triggered by any kind of presentation of X or any event or memory that in any way involves X, it seems reasonable to claim that that animal has a concept of X that, in and of itself, does not differ significantly from the kind of concept of X that humans have.

Evidence against Advanced Cognition

However, as shown earlier, the case for advanced cognition in prelinguistic organisms is based on indirect evidence: the capacity of such organisms to behave in ways that strongly suggest, but do not necessarily prove, their possession of human-like concepts. In drawing conclusions to such behaviors we should always heed the caveat of Shettleworth (2010: 4): "Research often reveals that simple processes apparently quite unlike explicit reasoning are doing surprisingly complex jobs."

At least as significant as the things nonhumans can do are the things nonhumans *can't* do. If nonhumans do have advanced cognition, how is one to explain the list of uniquely human cognitive behaviors listed in Penn et al. (2008: 109) and cited in Chapter 1? "Human animals—and no other—build fires and wheels, diagnose each other's illnesses, communicate using symbols, navigate with maps, risk their lives for ideals, collaborate with each other, explain the world in terms of hypothetical causes, punish strangers for breaking rules, imagine impossible scenarios, and teach each other how to do all of the above." Surely the only thing that enables these behaviors is a form of cognition far in advance of what other species have. Mere lack of labels for concepts or of channels for utterance does not seem to explain nonhuman inability to devise such things. Concepts as abstract as "cause" might have been hard to deal with without words, but what about concepts of things like "fire"? If use of fire, as is sometimes suggested, began with the collection of smoldering material from lightning strikes, this did not require language. Nor did snares (made from lianas), deadfalls, or covered spiked pits, any or all of which would have greatly enhanced primate capacity for trapping game. If apes are so intelligent, how is it that no ape has ever produced any innovation along these lines?

Penn et al. (2008) attribute such nonhuman limitations to an inability to perform abstract reasoning in areas such as transitive inference and causal relations. They contend that even some simpler operations such as same-different and match-to-sample tests, operations that many species seem to master easily, depend on objects in the tests having perceptual properties and claim that if role-based (functional) properties are involved, the operations fail. Conversely "a human subject is perfectly capable of reasoning about a role-based category such as 'lovers' or 'mothers' or 'tools' without there being any set of perceptual features that all lovers, mothers, or tools have in common" (125).

This seems to be setting the bar too high. "Lovers," "mothers," and "tools" cannot exist as categories prior to language. Neither a mother, a lover, nor a tool has any perceptual properties that would enable a given entity to be identified as a lover, a mother, or a tool. To reason about such categories one has to have a set of predicates that define functions or relationships (e.g., "mother" is a female person who has given birth to someone) and another set that assigns class membership (e.g., "A hammer is a tool," "A saw is a tool," and so on). It is unsurprising that nonhumans lack such capacities. This follows straightforwardly from the fact that nonhumans don't have language.

What is really at issue here is the status of far simpler concepts, concepts that could have a purely perceptual base. Consider the concepts that would be required for, say, the invention of a deadfall. All one needs to do is to imagine a forest trail and a log propped over it supported by slender sticks in such a way that if one stick is dislodged, the log falls; then imagine an animal coming down the trail and dislodging the stick. Everything involved in this scene is a concrete entity with perceptual properties. Chimpanzees love meat when they can get it. If they have even the most elementary concepts, why haven't they thought of anything like this?

Resolving the Paradox of Cognition

Nonhumans must have advanced cognition and human-like concepts because there are so many things they can do. Nonhumans cannot have advanced cognition and human-like concepts because there are so many things they *can't* do. This paradox must be resolved if we are to understand how humans relate to other species.

A good place to begin is the difference between online and offline thinking (Bickerton 1990b). Online thinking occurs when an individual is involved in some specific activity, and that activity is the focus of thought. A bat finding its way around a deep cave, a hawk diving on a fast-moving rabbit, or a human driving a complex route through heavy traffic are all engaged in online thinking. Offline thinking occurs when the topic of thought has nothing to do with the thinker's current behavior—a human at rest, mentally designing a deadfall, for example. Are nonhumans capable of offline thinking? Since for any organism without language offline thinking would necessarily be a private operation, we cannot say for certain that they are not. However, there is no good reason to assume that they are. In our case, the ability to think offline is what underlies the immense creativity and behavioral variability of humans. The absence of that ability in all other animals would suffice to account for their minimal creativity and minimally variable behavior. Similarly the ability of humans to solve problems by performing abstract logical operations and the inability of nonhumans to solve similar problems unless these concern known objects with physical properties inevitably suggest that offline thinking is restricted to humans.

But attributing cognitive limitations to an inability to perform offline thinking merely restates the problem in other words. Why can't nonhumans do offline thinking? Until we can find out how concepts and categories are actually represented in human and nonhuman brains respectively, no one can do more than speculate. Under these circumstances, the best strategy is to assume the minimal amount of difference between human and nonhuman concepts that is consistent with the massive observable differences between the behaviors of humans and nonhumans.

Assume, as so many already have, that there is no substantive difference between human and non-human concepts. What could then account for the behavioral differences? Suppose that despite that similarity, nonhuman concepts could not be freely and reliably linked with one another, and could not even be voluntarily accessed. This would effectively prevent off-line thinking; nothing that the organism was actually engaged in doing would trigger the necessary concepts. If the assumptions here are correct, there is nothing more to be said. If they prove incorrect, this can only be because we have overestimated the quality of nonhuman concepts. We would then have to think about what quality of nonhuman concepts could make them unlinkable and inaccessible. But we would not be forced to backtrack on our original assumption, because the problems of linkage and accessibility would remain.

The assumption that animals capable of complex behaviors have concepts that differ little if at all from human concepts in their essential makeup is supported by some recent studies (e.g., Newen and Bartels 2007; Aguilera 2011; Chittka and Jensen 2011). Thus differences in the scope of human cognition might be accountable not in terms of the constitution of the concepts themselves but with factors such as the possibility of voluntary retrieval of concepts and/or the existence of strong neural links between areas where concepts are stored.

To perform offline thinking both voluntary retrieval and neural linkage are essential. Any concept has to be continuously accessible and immediately retrievable if any offline thinking is to be done. But nothing is gained by retrieving individual concepts in isolation from one another. It is essential to be able to link potentially any concept with any other concept, or at a minimum to have rapid linkage between a wide range of concepts so that invoking one will immediately give access to others related to it. It can hardly be an accident that these prerequisites, as well as being basic essentials for any complex thinking, are identical with those required for conducting fluent linguistic communication.

From Animal Communication to Protolanguage 81

But how could the prerequisites themselves have come about? We return to this topic below.

The Initial Step toward Language

If we want to start from the earliest stage of language, and if at the same time we want to begin by exploring the likeliest possibility—that behavior caused changes in the mind/brain rather than vice versa—then we have to look at prelinguistic communication, unpromising as that may initially seem. (Compare the treatment of this issue in Bickerton 1990b with that in Bickerton 2009, which Balari and Lorenzo 2010 treated almost as an act of religious apostasy from the anticommunicative position of the earlier work.) Only if this course proves fruitless should we look at developmental or other organism-internal phenomena. It is all too easy to tell just-so stories about hypothetical internal developments—serendipitous mutations or still mysterious "laws of form" that might have led to language. But paying lip service to biology while cherry-picking which biologists and which biological phenomena to pay heed to (the approach unfortunately adopted by too many self-described "biolinguists") seems relatively unlikely to yield fruitful results. It is important to note that nothing in the present treatment entails or even implies that "language developed from prior communication," the bugbear that so frightens Balari and Lorenzo (2010) and others who subscribe to "biolinguistics" (see Jenkins 2000; Boeckx and Grohmann 2007). What evolved from prior communication was not language or anything like language, but rather an enhanced communicative repertoire that opened the gates for the subsequent development of language.

As Conway Morris (2003) points out, convergence (adoption of similar solutions to similar problems) is ubiquitous in evolution. This fact suggests a modus operandi for an origins-of-language search. If any other species, however remote, has ever exceeded even one of the limitations that characterize the rest of nonhuman communication, then the problem(s) that occasioned such a development may have been faced by human ancestors too. Should that prove to be the case, the first plausible scenario for the first step toward language should be within reach.

The Neglected Hymenoptera

What is perhaps the only clear breach of prior communicational limits is one that involves displacement. Displacement is listed among the design features of language (Hockett and Altmann 1968) and defined as "the ability to talk about things that are not physically present." The only species for which mechanisms of displacement are known are phylogenetically remote from humans (bees and ants); the overall process can be inferred in the case of ravens (Heinrich 1991), but here the method of execution is still unknown. Bees and ants employ very different mechanisms from one another. (On complex dances in bees, see Frisch 1967; Gould 1976; Dyer and Gould 1983; On chemical or behavioral methods in ants, see Wilson 1962; Holldobler 1971, 1978; Moglich and Holldobler 1975.) The behavior itself and its function, however, are identical

The phyletic distance between humans and hymenoptera, the fact that the latter operate by instinct (recall that humans are supposed to operate with "advanced cognition"), the profound differences in organization (social vs. eusocial), and the massive differences in brain size have so far discouraged language-evolution researchers from seriously considering the possible relevance of hymenopteran communication. But before writing off bees and ants there are certain facts that make such differences far less significant. First, language *as a brain-internal physical mechanism* lies far beneath the reach of conscious cognition and is consequently as much an instinct as anything bees or ants do. Second, some scholars have quite seriously proposed that, abstracting away from reproductive behavior, humans should be included among eusocial species (Foster and Ratnieks 2005), even if "only loosely" (Nowak et al. 2010). Third, and most important, work on evolutionary convergence (Conway Morris 2003, 2008; Liu et al. 2010; McGhee 2011) has shown that factors like relative brain size and phyletic distance are quite irrelevant to convergence processes. It would be perfectly possible (and highly desirable) to create a convergence map of organisms, similar in principle to cladistic diagrams, but instead of grouping species by their physical features, grouping them by their habitats and means of obtaining subsistence.

When we recall that large-scale cooperation also is found only among hymenoptera and humans, the legitimacy of comparing these apparently widely different organisms becomes even more apparent. We need to look at the problems faced by ants and bees that led to the relevant communicative and cooperative behaviors to see whether in fact similar problems were faced by human ancestors (but not by any other primate species).

Both bees and ants are extractive foragers (omnivorous ones, in the case of ants). Both exploit food sources that are often large and relatively short-lived (patches of flowering plants in the case of bees, dead organisms in the case of ants) and that could not be fully exploited by lone individuals. These factors make it necessary to recruit nest mates by imparting information about the whereabouts and in some cases the nature or quality of the food sources. The fact that the latter are normally at some distance from where the information is transmitted forces displaced communication.

A very similar ecological problem was faced by human ancestors, but by no other primate species, around two million years ago. It involved confrontational scavenging, in particular the scavenging of megafauna carcasses. Some historical background is required to show how this came about.

A Scavenging Niche for *Homo*?

Over the period from four million to two million years ago, the climate of East Africa became progressively drier. Consequently *"Homo* is the first hominid to exist in areas of fairly open, arid grassland" (Reed 1997: 289). During this period, great changes took place in the distribution of species, as shown by statistics from the Omo Valley, a habitat central to the area of most numerous hominid sites (Bobe et al. 2002). Primates, their regular food sources severely reduced, fell by nearly 50 percent in the million years between three million and two million years ago, while bovids, taking advantage of large open grasslands, increased their representation in the total fauna more than threefold: from 17 to 54 percent. These increased numbers meant that meat became the food source with the highest calorific yield for the least expense of energy and should therefore, according to optimal foraging theory, have been preferred by any species able to exploit it (Stephens and Krebs 1986).

Meat had presumably always been a welcome addition to primate diet. (For discussion of hunting by chimpanzees, see Boesch 1994; Boesch and Boesch 1989; Mitani and Watts 1999.) Note, however, that the chimpanzee's most favored hunting strategy—isolating individual monkeys in tall trees—would have seldom been available under savanna conditions, while the technology of *habilis* or early *erectus* would have been inadequate for the capture of large or even medium-size prey (Binford 1985). No doubt some hunting was undertaken, but it would have been opportunistic, limited to smaller, sick, or aged animals, and inadequate to make up for the loss of frugivorous resources. Human technology, however, was extremely well adapted for scavenging; it included sharp stone flakes capable of cutting through the thickest hide (Schick and Toth 1993) and hand axes that, in addition to their use in butchery, had aerodynamic properties that would have made them effective projectiles (O'Brien 1981) for driving off rival scavengers.

Among other primates, scavenging behavior is rare or nonexistent (Teleki 1975, 1981). This has been taken as evidence against scavenging by human ancestors (e.g., Geist 1987). However, what food a species obtains and how it is obtained are determined less by genetic factors (except insofar as these may make certain foods indigestible) than by what the environment offers. The habitats of other primates made scavenging unnecessary, but some habitats of human ancestors made it unavoidable if those ancestors were to survive.

In any case, evidence that some varieties of *Homo* engaged in scavenging is abundant. Prior to two million years ago, cut marks on bones (indicative of human butchery) are normally found superimposed on tooth marks of other species, indicating that prehumans had accessed bones only after other animals had partially or wholly defleshed them. After two million years ago, tooth marks are with increasing frequency found superimposed on cut marks, indicating that prehumans had accessed the bones first (Bunn and Kroll 1986; Blumenschine 1987; Blumenschine et al. 1994; Monahan 1996; Dominguez-Rodrigo et al. 2005). Since some of the animals concerned are too large to have been successfully hunted with the armament of contemporary hominids, scavenging becomes inescapable.

Significantly, other factors from quite different areas point to changes in human behavior and physiology also arising at approximately two million years ago. Two that seem obviously related are changes in foraging patterns (Larick and Ciochon 1996) and changes in physiology suggesting an adaptation for endurance running (Bramble and Lieberman 2004). The first of these involves a change from catchment scavenging (where hominids worked a restricted area around a central place) and territory scavenging (where hominids ranged over an extended area with no particular focal spot). The second involves differences between *habilis* and *erectus,* the latter being larger and taller, with longer legs and an essentially modern skeleton. Both these changes are consistent with a niche that, while not limited to scavenging, significantly involved the location and procurement of large carcasses over what were probably quite wide areas.

However, the type of scavenging involved was inevitably confrontational. Since prehumans lacked the technology to kill larger mammals, they had to take carcasses of animals that had either died natural deaths (infrequent for any but the largest species) or that other carnivores had already killed. In either case, they were likely to encounter fierce competition from both predators (who also scavenge when possible, since scavenging rather than hunting conserves energy) and other scavengers. In fact even into *erectus* times, humans were as likely to become prey as to take prey (Hart and Sussman 2005). Only well after two million years ago was there a significant decrease in the number of human bones found in caves used by predators around Swartkranz, South Africa (Pickering et al. 2008). Lacking both artificial and natural (teeth, claws, etc.) weapons, prehumans were left with a single resource: numbers. Only if they were able to recruit numbers large enough to drive away competitors could they hope to gain first access to most carcasses.

In other words, the ecological problem they faced was strikingly similar to that which led to the emergence of displacement among ants and bees, the only difference being that the perishability of the food source was now accompanied by the need to exclude competitors (nonexistent for bees and relatively trivial for ants, though extremely important for ravens). Both factors made speed of execution essential. But recruitment, in a non-eusocial species, inevitably presents problems.

Recruitment requires cooperation, and primates seldom cooperate. On the relatively rare occasions when they do (which usually involve sharing of meat as a symbol of value), cooperation can be accounted for in terms of kin selection (Hamilton 1964), reciprocal altruism (Trivers 1971), "tolerated scrounging" (Isaac 1978), the consequence of sanctions against cheaters (Clutton-Brock and Parker 1995), or "costly signaling" for the sake of some consequent benefit (Hawkes 1991). But even when cooperation occurs, it rarely extends beyond dyads. The kind of cooperation required in recruitment for carcass seizing, a kind that necessarily involves quite large groups of individuals, is wholly unknown in all living primate species but one.

De la Torre Sainz and Dominguez-Rodrigo (1998) place the origins of human cooperation squarely in the context of meat consumption, although since they cannot decide on the relative roles of hunting and scavenging their proposal is less specific than that developed here. Without some such recruitment scenario, there is no convincing account of how human cooperation (of a strength unparalleled outside the hymenoptera) got launched. Cooperation-competition conflicts have been modeled extensively in recent years through the medium of game theory (Maynard Smith 1982; Axelrod 1984). No currently available model exactly represents all of the dimensions involved in confrontational scavenging scenarios, but nonlinear N-person public good games (Archetti and Scheuring 2011) come close: "In these models cooperators pay a cost and contribute to a public good that, in the simplest case, can be consumed by all members" (Bickerton and Szathmary 2011). In such situations, the cost (in this case, time and energy possibly wasted, plus some risk of injury or death in confrontations with rival scavengers) should not outweigh the public good (a large supply of meat that might sustain cooperators for several days). If all cooperated, everyone would receive this benefit. If an insufficient number responded, no one would receive any benefit. The latter factor would make punishment of defectors—a powerful factor in encouraging cooperation, according to Boyd et al. (2003)—much more likely.

However, the choice between cooperation and defection depends crucially on information—information of a kind no other primate, and hardly any other species, was equipped to give, since the source of that information would in most cases lie well outside the sensory range of potential recipients.

Confrontational Scavenging and Displacement

Consider first the constraints that determine unit size in foraging. While larger numbers are more effective for protection from predators, animals must forage over an area that can meet their energetic and nutritional requirements (Chapman et al. 1995), making overly large foraging bands an uneconomic proposition. Consequently fission-fusion foraging patterns have been observed for several primate species (Kummer 1971; Lehmann et al. 2007), and in savanna environments, "daytime food searching demands required individuals to forage in small parties" (Aureli et al. 2008: 31). However effective larger bands might be as protection from predators, such bands foraging as units could not have covered sufficient ground to feed all members, and the practice would soon have led to starvation. But now suppose one of these small parties encountered a large, freshly dead carcass. What was it to do? If other scavengers were already present, small numbers could not safely and simultaneously butcher the carcass and keep the other scavengers off; some members of the party would likely themselves end up as prey. They would need the help of the other parties that composed their band. But in all likelihood, those other parties would be out of both sight and earshot, perhaps miles away. How could information rich and precise enough to ensure effective cooperation be transmitted? Only by extending existing modes of communication through the addition of displacement.

Note that this situation both is unique to human ancestors and links a fundamental (nutritional) need with a behavior that requires something beyond normal animal communication for its successful execution—two factors that previous theories of language evolution conspicuously lack. How was that execution carried out? We may never know, but if ants and bees could find a way, surely an advanced primate could. Perhaps something like the tandem running of ants (where one ant literally grabs another and drags it along) was involved. Probably individuals imitated sounds or actions typical of the carcass's species, since this would give an idea of the carcass's size, which in turn might encourage cooperation. Distance and direction would help too, as with bees. Clearly in such a situation whatever worked communicatively would be adopted, making irrelevant all the frequently voiced arguments over whether "the earliest forms of language" were vocal or manual.

Note the existence of yet a further factor that, while of limited significance at the time, would turn out to be indispensable for the evolution of both language and advanced cognition. Hitherto (unless we accept the dubious premise that functional-reference signals were proto-words for predator species) no signal outside the hymenopteran repertoire conveyed specific reference to objective aspects of, and entities within, the natural world. In other words, the notion that the world might consist of nameable objects was literally inconceivable to animal minds. This is another problem that researchers in the field have consistently underestimated.

Recruitment for confrontational scavenging forced prehuman minds to accept the notion. When a prehuman who had just discovered a carcass appeared before conspecifics gesturing and making strange sounds and motions, the motive could only be informational: "There's a dead mammoth just over the hill, come help get it!" Recall that some ant species present samples of the food available to encourage cooperation. Our hypothetical prehuman could hardly bring along a chunk of mammoth meat, but he or she could convey, by mimicry of the motions and/or sounds made by the animal in question, what species that animal belonged to. The ability to name the species would prove not only central to language when language finally emerged but would also establish the linkage between voluntary signals and their related concepts crucial for the development of both language and advanced cognition.

In Bickerton (2009), only the scavenging of megafauna—carcasses of ancestral elephants, rhinoceroses, and the like—was considered. Bickerton and Szathmary (2011) propose a more gradual development of confrontational scavenging, starting with carcasses of smaller animals either abandoned or attended by only one or two predators or a few vultures, progressively graduating to larger and better guarded carcasses as the ability to drive off competing scavengers (by improved distance and accuracy of missile throwing) and to butcher carcasses (with Acheulean rather than Oldowan technology) both improved. It has been suggested (Robert Blust, personal communication, 02/12/13) that contemporary hominids could have used fire as well as throwing to deter competition. This presupposes that such hominids could actually have made fire. (They could hardly have maintained fire over long periods just in case they encountered a large carcass.) However, the process of stone knapping would sometimes produce sparks, which might occasionally ignite dry vegetable material, so this possibility for the beginning of controlled fire and its first use cannot be ruled out.

In any case, there is evidence for increasing meat consumption starting around two million years ago (Eaton et al. 2002; Ungar et al. 2006). After the carcass was seized and butchered, selected portions would presumably have been carried to the nearest relatively secure place and consumed there. It is even possible that hominids proactively tracked herds of large herbivores to watch for sick animals that might form future prey, though on present knowledge this must remain speculative.

Any theory that seeks to explain why and how language originated must be able to pass a series of tests. The selective pressure involved must be one unique to human ancestors; otherwise we have to explain why the same pressure in other primates had no effect. The result of the process must be fully functional from the beginning; otherwise no form of language would ever have established itself. Since a handful of signals featuring displacement does not constitute a significant expenditure of energy, those signals are "cheap tokens" (Zahavi 1975, 1977) likely to be disbelieved and disregarded by members of a species capable of tactical deception. Since primates are selfish, information exchanged must be equally useful to sender and receiver (just as cooperation must benefit both inviter and invitee).

Only the theory presented here passes all four of these tests. Uniqueness is satisfied by the fact that confrontational scavenging was not undertaken by any other primate. Initial functionality is satisfied by the fact that the first small handful of signals would have brought tangible and immediate benefits. The "cheap token" credibility problem is resolved by the fact that the presence of carcasses would have quickly confirmed statements of their availability (or, in case of disconfirmation, would have quickly led to sanctions against liars). Selfishness would not have been an obstacle, since without an exchange of information, neither sender nor receiver would have received benefits.

Scavenging and Wallace's Problem

It is essential to emphasize that if evolution had stopped at the point we have now reached, Wallace would not have had a problem. In Chapter 1 we saw how Wallace was disturbed by the fact that language and human cognition were far in excess of the needs of any prehuman species—so deeply disturbed that he could not believe normal evolutionary forces could have given rise to them. For that reason alone he felt obliged to make a divine exception for the intellectual powers of humans, even at the cost, as Darwin pointed out, of assassinating his own brainchild.

What human ancestors now had was an enhanced communication system adequate for the development of a new niche: the confrontational scavenging niche. Wallace would have been happy with that. Why would those ancestors have needed anything more? What would have selected for more complex adaptations?

Language in its modern form is so clearly adaptive, provides such an unprecedented degree of control over the environment, that we automatically assume that once it had begun, any improvement in it would have been selected for. Two different but complementary considerations suggest that this assumption is at best premature. First is the fact that, though language in its modern state may be highly adaptive, it is far from obvious why intermediate stages, especially the earlier ones, would in themselves have been adaptive. We can tell one another just-so stories about why they might have been. We can suggest, for instance, that individuals with better language skills would have acquired more mates or achieved more positions of leadership, thereby increasing their contribution to the gene pool. But these are not even just-so stories so much as Hail Mary passes to get around Wallace's problem and show that run-of-the-mill natural selection *can* account for every step in human evolution.

The second consideration involves the paleoarchaeological record. If language had yielded significant results in developing new and more efficient behaviors, we would expect to see starting around two million years ago a gradual but steady improvement in artifacts and other physical traces of human behavior. In fact we see nothing of the kind. Tool inventories hardly increased over most of that nearly two-million-year period. Even existing tools changed at a glacial pace: the Acheulean hand ax retained the same shape and proportions for over a million years. "The Oldowan and Acheulean industrial complexes are remarkable for their slow pace of progress between 2.5 and 0.3 [million years ago] and for limited mobility and regional interaction" (Ambrose 2001:

1752). Such a sluggish pace suggests that, *pace* Pinker and Bloom (1990), language was not being strongly selected for.

This same argument could, of course, be used for claiming that language in any shape or form simply didn't exist until fifty to a hundred thousand years ago. Unfortunately to do so you have to believe one of two things. You have to believe that language with all today's complexities blossomed from an alingual state without any triggering circumstance as a result of yet unknown laws or serendipitous mutations. Alternatively you have to believe that everything needed for language was in place already save perhaps one thing—recursion, say—which for some similarly mysterious reason and by equally mysterious means managed to wrap all those prerequisites into one exquisitely functional package.

Note that the adoption of such beliefs doesn't in any way solve Wallace's problem. Since the state of affairs immediately preceding the appearance of anatomically modern humans had (as noted earlier) changed very little in the preceding two million years, the gap between human needs and human capacities had scarcely narrowed. Much more limited means, both linguistic and cognitive, would have sufficed for life as it was lived one or two hundred millennia ago. And for no other species did an abrupt and unaccountable event provide a cognitive (or any other kind of) increment that was so much more than nature could have needed.

If we rule out normal selective processes and magical versions of evo-devo, what's left? The answer is just two things: words and their neural consequences. The invention—for it can only have been an invention, albeit not a conscious or even an intentional one—of symbolic units had crucial consequences for the brain, giving it material to deal with of a kind it had never experienced before. The ramifications of this theme will carry us through most of this chapter and the next. The place to start is with displacement. This offers a way, perhaps the only way, through which symbolism could be attained.

Why Displacement Matters

Choice of displacement as a road into language may surprise some readers. Displacement is not what most people think of first when they consider the most clearly distinctive properties of language. Arbitrariness is a common choice and is indeed ubiquitous in language. Apart from a handful of onomatopoeic items, words bear no relation to the objects they symbolize, as illustrated by many familiar examples (e.g., representation of the same domesticated animal in contiguous languages by Spanish *perro,* French *chien,* German *hund,* and English *dog*). The dog is an interesting example, because one of its attributes not only lends itself to onomatopoeia but has actually yielded an onomatopoeic expression: *bow-wow.* But *bow-wow* never graduated out of baby talk or motherese, and *bark* itself is rendered in other languages by equally nononomatopoeic terms: *ladrido* in Spanish, *aboiement* in French, *bellen* in German.

But arbitrariness is not unique to human communication. For instance, the signals given by vervet monkeys and chickens for the appearance of aerial predators bear no resemblance either to one another or to anything in the sounds or appearance of hawks or eagles. Moreover abandoning arbitrariness altogether would still leave language intact; *the bow-wow goes bow-wow* obeys all the relevant nonlexical rules of English.

Symbolism (cf. Deacon 1997) would make a better choice as *the* distinctive language characteristic. As opposed to units of animal communication, which refer (if they can be said to refer at all) directly to particular objects or events in the real world, words refer to things via mental concepts of general classes of entity and behavior (Saussure 1959). *Dog* does not refer to any particular dog, or *run* to any particular act of running. Indeed reference to a particular dog can be made only by adding to *dog* some specific identifying information. In other words, symbolism entails a concatenation of units, just as animal communication entails the absence of concatenation. The question is: How can symbolism be attained? Unlike displacement, symbolism has no precedent in any other organism; humans really are the only "symbolic species." Yet I shall argue here that displacement forms perhaps the only route to symbolism.

First, following Deacon's (1997) thorough treatment, it is necessary to take account of the various classes of signal. The first to deal systematically with these was Peirce (Buchler 1955), whose "original distinction between icons, indices, and symbols is based on the fact that an 'icon' has physical resemblance with the object it refers to, an 'index' is associated in time/space with an object, and a 'symbol' is based on a social convention or implicit agreement" (Cangelosi 2001: 93). But for the kind of symbolism found in language, more than this is involved. Cangelosi, like many philosophers and psychologists, repeatedly commits himself to the belief that linguistic symbols refer directly to real-world objects rather than via our concepts of those objects. For instance, in comparing the reference of vervet alarm calls with that of words, he states that "symbolic associations have double references, *one between the symbol and the object,* and the second between the symbol itself and other symbols" (94; emphasis added). This alone would cast doubt on his claim to have arrived at symbolism through a series of computer modeling experiments.

Cangelosi's (2001) virtual "organisms" come already equipped with tools to build a vocabulary—in this case, a binary code consisting of a finite (and small) number of bits. The organisms construct a series of possible words by permutations of 0 and 1. They are then tasked with finding (by trial and error) a distinction between poisonous and harmless mushrooms and eventually home in on an agreed designation for each class. A subsequent experiment by Cangelosi is supposed to show a progression from symbolizing specific features to symbolizing whole categories. But these experiments would not have worked without the prior assumption that the virtual organisms had at least the notion that signs could represent abstract classes or categories of things rather than particular individuals. It seems obvious to humans that this should be so, because from their first to their last breaths all humans inhabit a world drenched in symbolism. This could hardly be otherwise for any species with language, which is symbolic throughout.

Consequently the notion that symbolic representation is somehow natural and doesn't require any special explanation is very hard for us to get rid of. The tacit belief that other animals see the same world as we do, filled with potentially nameable objects and events for which they simply haven't yet found names, probably goes a long way toward explaining the frequently expressed belief that a mere increase in brain size or intelligence or social complexity or understanding of others' intentions (Theory of Mind) would have sufficed to launch language. But nothing in ethology or ecology or biology or comparative psychology supports the notion that for other animals the world is composed of a set of discrete categories that are potentially nameable. If a concept representing some animal species X can surface in the mind only under one of two conditions—either when some physical manifestation of X or something closely connected with X triggers it or when it is accidentally excited by random activity in the brain—there is no way any nonhuman animal could even conceive of the possibility that an arbitrary chunk of sound or a gesture could stand as the permanent label of some discrete class of entity.

The distinction between the "namelessness" of all other species' *Umwelts* and the "namefulness" of the human *Umwelt* is too absolute for a transition to have been made without intervening steps (a factor not considered in Deacon 1997, where a direct emergence of symbolism from specific human behaviors is envisioned). Only displacement could provide those steps.

For alingual prehumans, communicative signals could only have been iconic (like baring the teeth as an indication you were prepared to bite) or indexical (like a predator warning indicating some immediate threat). These signals by their very nature can relate only to phenomena that are physically present. To start with, it was revolution enough to introduce displacement. Displacement breached the barrier of the here and now, but the things it indicated were still particular real-world entities rather than conceptual classes. And as noted, only a tiny handful of species ever achieved this, and only when obliged to do so by pressing ecological needs.

If there were only iconic or indexical signals, how could displacement have been achieved? Again, we need to look at how bees and ants do it. (Holldobler and Wilson 2009 represent what is probably the best recent overview.) Ant signals are relatively straightforward: presentation of food samples is iconic, while tandem running and chemical trails are indexical. Honeybee signals are quite complex, and while these are believed by some to have evolved from simpler signals in other bee species such as bumblebees (Dornhaus and Chittka 1999), there is much more one would like to know about their evolution, since their relation to their referents is more abstract (hence more symbol-like) than anything else outside language. (See Dyer 2002 for a review.) Some seem to mix both iconic and indexical features, such as the waggle dance that indicates relative distance with the length of its straight run (iconic) and direction with relation of its axis to the sun (indexical). Since dances take place where the sun is invisible, the latter attribute is explicable only if modern bees descend from a species that danced in the open (Dyer 2002). We can thus assume a mix of iconic and indexical signals used by the first confrontationally scavenging hominids.

The rarity of iconic and indexical items in language today is, of course, no indication whatsoever of their frequency earlier stages of development. The whole point about displacement is that any type of signal used when no referent for that signal is currently perceptible to any sense has already taken on what is perhaps the most critical property of symbolism: its ability to evoke phenomena beyond a recipient's sensory range.

What could those early signals have conveyed? Better, what would have best achieved the purpose of recruitment? As with ants, some notion of what kind of food was to be expected would be helpful, and here the nature of the carcass could be indicated by imitating characteristic sounds or postures of the species concerned. If members of the recruiting group were returning immediately to the carcass, something like tandem running would have been helpful. If they were not, then direction must have been given, and distance—at least "near" as opposed to "far"—would have been an encouragement. Characteristic sounds or gestures might have been associated with notions like "Hurry!" or "Tell others."

With regard to vocabulary extension, it is worth noting that some very recent work in robotics supports the claim that, once a tiny starter vocabulary had appeared, words relevant to the behaviors associated with that vocabulary would have increased naturally. Schulz et al. (2011) and Heath et al. (2012) report on tiny robots ("lingodroids") that, equipped with the power to utter syllables and communicate them to each other, worked out their own vocabulary for divisions of time and distance. But of course, even with such extensions, the emerging protolanguage would have been little more sophisticated than the ant and bee "languages" that have never expanded for use outside their original foraging contexts. Indeed it seems likely that for some quite considerable time, what has been described here was all there was—a purely functional medium, exactly like the displaced communication of ants and bees, tied in this case to the specific context of confrontational scavenging.

Even if a sound like the trumpeting of a mammoth was used, this would indicate not mammoths in general (as the modern word would) but merely "one particular dead mammoth," the one that was there right then. But in a species with a brain even the size of an ape's, things were unlikely to stop that way for ever.

In accounting for phenomena, evolutionary convergence and constructed-niche resemblances can take us only so far. Species occupying similar niches and adopting similar solutions to the problems those niches present will still vary in their subsequent behavior according to the differing capacities they bring to the niche. Prehumans, like other primates, had individual personalities, hence a fairly wide range of behavioral variation; bees and ants do not. Prehumans had concepts (of some kind), which ants and bees presumably did not. Prehumans had brains that were several orders of magnitude larger than the brains of ants or bees. Therefore the ultimate consequences of adopting a protolinguistic trait like displacement could hardly be the same for prehumans as they were for bees or ants.

Consider the many causative actors that have been proposed for the origin of language: tool making, social maneuvering, female choice, control of infants, and so on, all things of no concern to bees or ants. While none of these sufficed to *start* language, each would have eventually contributed to *developing* language once the initial steps had been made. Once it had been grasped that a sound or gesture could represent entities or occurrences outside the sensory range, that notion would surely have been put to work in a variety of ways. Pedagogy is one: warning offspring of dangers that were not immediately present but that might drastically impact their survival chances in the future. You can tell any number of just-so stories along these lines: a father, for instance, pointing to a particular kind of footprint and making the sign for "tiger," whatever that was. Other motivations may have helped to expand the vocabulary, such as gaining prestige by being the first to draw attention to some novel event or circumstance (Dessalles 2000, 2009), a possibility Desalles (2010) suggests may be an evolutionarily stable strategy in the sense of Maynard Smith and Price (1973). However, any account of how the earliest vocabulary developed can only be hypothetical.

From Prehuman to Human Cognition

We are on slightly firmer ground when we consider the consequences a developing protolanguage would have had for the prehuman brain. Earlier I suggested that there might be no difference between human and nonhuman concepts per se, yet serious differences in cognitive behavior might still arise from differences in the extent to which concepts could be retrieved and linked to one another. Let's start by considering how concepts are instantiated in human brains.

Concepts (human or nonhuman) must be based on a distributed array of percept-based data, but are they stored *only* as a distributed array? For concreteness, take the concept "leopard." This might merely consist of a broadly scattered collection of leopard characteristics, with their shapes and colors in areas of the brain devoted to visual processing, their sounds in areas devoted to auditory processing, and so on (Allport 1985; Saffran and Schwartz 1994; Martin 2007). However, if this is all there is, it becomes difficult to explain how higher-order generalizations across concepts (making links between concepts that may have only one or two features in common) can be attained. This problem can be resolved by Damasio's (1989; Damasio et al. 1996) notion of "convergence zones," additional brain areas where concepts with similar sensory or functional characteristics are stored. It has also been suggested that the distributed representation must be linked to some central place, or "hub" (Patterson et al. 2007), so that each whole concept has (in addition to its distributed perceptual base) a representation with a single locus, the loci themselves being grouped in terms of semantic similarity (Riddoch et al. 1988; Caramazza et al. 1990). In fact since the brain operates with some degree of redundancy, these theories are not mutually exclusive. There seems no obvious reason why concepts could not have both a hub where all the distributed percept-based representations come together and convergence zones where concepts with qualities in common are grouped.

According to one source (Chater and Hayes 1994), there exist no operational means for disconnecting conceptual knowledge from lexical knowledge, consequently no way to determine the nature of concepts for any alingual species, or even whether other species have any concepts, unlikely though the latter state might seem. Others regard the current status of the issue as due to neglect: "With the possible exception of developmental psychologists . . . there is typically little overt interest within cognitive science in comparative analyses of the similarities and differences in conceptual behavior between humans and other animals" (Zentall et al. 2008: 13). They claim their review of a wide range of experimental studies provides "good evidence that many species of animals are likely to have some of the same conceptual abilities that have typically been reserved exclusively for humans" (41).

Evidence from a quite different direction—studies of economy in brain wiring (e.g., Cherniak 2005; Cherniak et al. 2002)—supports the notion that animal concepts may even be structured exactly like human concepts. Imagine a brain that represented concepts in a distributed-only mode. Since one of the main functions of a brain is to ensure that its owner responds appropriately to changes in the environment, each of the percept-based representations of some aspect of the concept—the sound, smell, sight, movement, and so on of a leopard, say—would have to be wired to every other such representation in order to ensure that whatever stimulus was currently being perceived did originate with a leopard and not just something that from a certain angle or distance merely looked or sounded like a leopard. Moreover, in the absence of a hub, every one of these representations would have to be separately linked to the appropriate motor response to leopard recognition (flee, hide, climb a tree, wait and see, etc.). One can easily picture the criss-crossing cat's cradle of axons that would result. In contrast, wiring of each representation to a hub and from the hub to each response mechanism would sharply reduce both the multiplicity and the complexity of the wiring, as well as reducing reaction time in a highly survival-inducing fashion.

If nonhuman concepts are structured much as ours are, then, as noted, either concept retrieval or interconceptual linkage or both may be missing, thereby accounting for cognitive disparities. Assume this to be the case. Why would words change this picture? Recall a crucial quotation from Chapter 1: "If it be maintained that certain powers, such as self-consciousness, abstraction etc., are peculiar to man, it may well be that these are the incidental results of other highly-advanced intellectual faculties, and that these again are mainly the result of the *continued use* of a highly developed language" (Darwin 1871: 101, emphasis added).

That "continued use" must have been inspired, for I doubt whether, given the state of knowledge 140 years ago, Darwin could have dreamed of explaining how language might have had such an effect. However, it may now be possible to support his insight.

Let's look at how and why concepts can be triggered in the absence of symbolic signals. For concreteness, take a lion and some prey animal. The prey sees a lion, and its appearance triggers visual representations in the prey's brain, which in turn trigger lion recognition, or more specifically "lion, at X distance, moving tangentially, looking uninterested." The neurons that constitute the "lion" hub fire at a sufficient level of arousal and in turn trigger the motor neurons responsible for "wait and see" responses. The lion goes away. Presumably the concept goes away too. What could bring it back, bar another lion appearance? There is no need for the prey to evoke it in the absence of any actual lion. Random brain activity could of course accidentally trigger the lion concept, as presumably happens in dreams. But could the prey intentionally summon up the concept, as a human child from age three or so could, just to "think about" lions? Even if it could, why would evoking that concept trigger other concepts that together would constitute a coherent "train of thought" about lions—how to avoid them, where to find them, how if necessary you might even succeed in killing one, or whatever? It is not clear that any mechanism exists that would bring this about in nonhuman organisms.

What Words Can Do

Now consider words. Utterances of the first displacement signals would have occurred at various stages of the confrontational scavenging scenario and would at first have been as tightly linked to those stages as, say, the vervet "leopard" alarm signal is linked to actual appearances of leopards. But as suggested earlier, a gradually increasing number of uses in an increasing variety of contexts would progressively bleach the signal of its scavenging-specific associations, slowly (this could have taken countless millennia) converting it from context-bound signal to context-free symbolic word. As is apparent from everyday experience, and as has been repeatedly demonstrated in experimental neurology (e.g., Kroll and Curley 1988; Klinger et al. 2000), representations of words and their associated concepts are tightly linked. Utterance of a word may simultaneously evoke any or all of the semantic properties of that word: locations of lions (Africa, zoos), sounds (a roar), physical appearance (mane, tufted tail), and so on. Equally, mention of any of these properties may evoke the word *lion*. "Conceptual knowledge does not appear to be removed from its perceptual origins, nor from its linguistic origins" (Boroditsky and Prinz 2008: 112),

How would the brain have reacted to getting words in it? Words were not like anything the brain had had to handle before. Prior to words, brains had not had to interpret anything that came from inside themselves—a simple but surely significant fact that seems to have gone largely unnoted in the vast literature on brain and behavior. Hitherto the brain's job had been to "take information from the senses, analyze that information, and translate it into commands that get sent back to the muscles" (Marcus 2004: 11). "Senses" of course include proprioception: realizing one is hungry sends one in search of food. Now, under certain circumstances, this process would have to be blocked; you couldn't have motor responses appropriate to the appearance of lions occurring whenever the "lion" concept was evoked by the word for lion. Inhibition of some typical motor responses would be accompanied by changes in the flow of information; instead of information always traveling from exterior organs of sense into association areas and then onward to motor areas, it would sometimes travel from one association area to another, perhaps several others, without triggering any motor action or triggering action only in the organs of speech.

All of this would require extensive redirection of afferent and efferent wiring, as well as a consistent (and ultimately large) increase in overall wiring connections. These factors would in turn select for increasing brain size. It is surely significant that genuine brain size increase (i.e., increase that exceeds the normally constant scaling ratio between brain size and body mass) begins around the time that confrontational scavenging began (around two million years ago) and that the neocortex, home to association areas, increased proportionately more than any other brain region to a size 3.2 times greater than would be predicted by body mass (Passingham 1975; Schoenemann 2009). Note that although brain size increases may have been *made possible* by the adoption of cooking (Park et al. 2007), by a cooling system for the brain (Falk 1990), or by other causes, nothing of this nature explains *why* human brains increased in size. It is often overlooked that making something possible in evolution by no means guarantees that that thing will evolve. If it did, no species could ever be trained to perform any behavior that it did not already perform in the wild.

Note too that utterance of words (we may now safely begin to call them that) had to be volitional. Voluntary utterance was not a total novelty. It has been known at least since Seyfarth and Cheney (1990) that warning signals are under voluntary control. (For instance, animals are less likely to warn if no close kin is around and will suppress signals altogether if they are alone.) Since the need for a particular word might arise at any time, without requiring anything in the environment to first trigger neurons in sensory areas and start concept evocation, the neural representation of words had to be constantly accessible.

All of the above formed a necessary condition for offline, human-like thinking. However, it still fell short of a sufficient condition. If I can do no more than deliberately think of a lion when no lions are around, little if anything has been gained. For offline thinking to begin, concepts must have strong neural links with one another. Would they have been linked in nonhuman predecessors? The answer would seem to be no. Both "lion" and "leopard" would probably have been linked to "run," but there are rules of economy that would discourage the formation of additional connections. "Because connections in the brain, particularly long-range ones, are a stringently limited resource both in volume and in signal-propagation times, minimizing costs of required connections strongly drives nervous system anatomy" (Cherniak 1994: 94). One consequence of the limitation on concept linkage is that making any kind of generalization about the behavior of lions and leopards, one that might have distinct survival value, is rendered impossible.

Words could have created the necessary neural links. The most likely way for neural connections to form, or to further strengthen once formed, is still Hebb's Rule: "When an axon of cell A . . . repeatedly or persistently takes part in firing [cell B], some growth process or metabolic change takes place in one or both cells such that A's efficiency, as one of the cells firing B, is increased" (Hebb 2002: 62).

Put alongside this Darwin's dictum about "continued use of a well-developed language," as well as the nature of niche construction, in which genetically unsupported behaviors change the environment in ways that cause those behaviors to be actively selected for, and you can begin to see the first crude outlines of a machine for jump-starting advanced cognitive behavior. The oftener words were used together, the stronger the connections that would form between their associated concepts. The more different words were used, the wider the semantic networks that enable offline thinking would spread.

Note again that these proposals are based on the notion that significant evolutionary developments begin in changed behavior rather than organism-internal changes. The issue is simply one of aligning the evolution of human cognitive development with the evolution of other unique novelties. Did beavers grow webbed feet and only then adopt an aquatic lifestyle, or did they embark on an aquatic lifestyle and only as a result of this gradually acquire webbed feet? Did bats acquire echolocation skills and then begin to fly and hunt by night, or did they experiment with night flying and hunting (driven from day by predators, perhaps) and then gradually enhance whatever minimal skills in the way of making sounds and perceiving echoes they already possessed? The answer is already there in behaviors common among our own species. Persons afflicted by blindness quickly learn to use the tips of their sticks to produce sounds and to interpret the echoes of these sounds so as to determine the approximate position of obstacles. They fortunately do not have to wait on some genetic change to provide them with echolocating skills. Humans (just as bat ancestors must once have done) lack any genetic or otherwise built-in equipment for echolocation, but there can be little doubt that, if some new disease were to blind the entire species, the minimal preexisting human capacity for processing echoes and locating their sources would be strongly selected for, and specialized genetically transmitted mechanisms for these powers would appear, grow, and rapidly improve.

In other words, behavior drives adaptation rather than adaptation driving behavior. Reluctance to accept this otherwise general law in the case of offline thinking probably stems from the lingering specter of dualism. There is a tendency to think that adaptations like beaver feet and bat echolocation are "physical" and therefore differ from adaptations like offline thinking that are "mental"—even though bat echolocation necessarily includes a large computational element (Roverud 1993;

Neuweiler 2003), just as offline thinking necessarily requires physical structures and processes in the central nervous system. Offline thinking is just as much (or as little) "physical" as anything else.

Though what I am claiming here may sound like the hoary old claim that "thought is impossible without language," it is not. First of all, what existed at the stage I'm describing wasn't language. It was still no more than an enhanced communicative repertoire that, as its units became more truly symbolic, might begin to be describable as a form of protolanguage. Second, there are many kinds of thought for which language is not needed, including kinds that can be performed by all animals above a certain (quite low) level of intelligence as well as by humans. All I am claiming is that a certain kind of thought probably limited to humans—thought that manipulates concepts of classes rather than of individual entities, that can transcend experience to create genuinely novel configurations—needed some kind of overt objects (signals or words) in order to get started. Any approach that is dynamic and process-based rather than essentialist recognizes that what might be a crucial necessity in the early stages of a process may be dispensed with entirely at some later stage, like scaffolding that necessarily surrounds a partly built house. While words might have been necessary to trigger the inception of offline thinking, there is no reason to suppose that concepts could not subsequently develop hub-to-hub linkage without any mediation of words. Indeed since simultaneous firing of concept hubs follows automatically from simultaneous firing of neuron groups representing their associated words, it seems likely that direct concept-to-concept linkage (making words superfluous to thought) would have been inevitable. We can thus short-circuit the seemingly endless debate as to whether there is, or cannot be, a "language of thought" that, to varying degrees, mimics externalized language (see Aydede 2010 and sources listed therein). Offline thinking can (nowadays) be carried on with or without the use of language, except for a small set of syntactic universals that reflect not the syntax of any particular language (see discussion in Chapter 5) but a basic structure for all languages. However, without overt linguistic behavior at the very beginning, nothing approximating any "language of thought" or any kind of offline thinking could have come into existence.

I am only too well aware that what is written here is largely speculative and must remain so until much more research has been done.

However, as noted, the relationship between human and nonhuman concepts has been woefully underresearched, and accordingly any account, however hypothetical—however wrong, for that matter—is better than the present state of affairs. Hopefully the account given here will sufficiently irritate specialists in the field to make them try to disprove it. Then we might really learn something.

Protolanguage

Processes reviewed so far would have begun with the extension of a signal repertoire in the direction of displacement—something, at that stage, indistinguishable from the context-bound "languages" of bees or ants—and turned it over time into what we may legitimately call a protolanguage. Some scholars reject even the possibility of a protolanguage. Piattelli-Palmerini (2010: 160–161) states that "it is very hard to even *define* words in the absence of . . . syntactic criteria" (original emphasis) and asks how protolanguage could exist "once you strip words *as we know them* from their internal structure and their compositional valence" (emphasis added). Note the italicized phrase in the second quotation. It is quite unrealistic to suppose that, one to two million years ago, "words" could have been anything like the words of modern languages as used by adults. But Piatelli-Palmerini has to look no further than any early-stage pidgin (Bickerton 1981) to find things that behave exactly like protolanguage words, that is, that are not be combined into any grammatical structures. Modern adult-speaker words may be, as he states, "fully syntactic entities," but to assume that their remote ancestors had to be so is to commit to an essentialist view of nature wholly at odds with any kind of evolutionary thinking.

Taking such a position also obliges one to assume that either syntax preceded words or words and syntax emerged simultaneously. If syntax preceded words, it must have been used for something other than language—a proposal floated over a decade ago (Hauser et al. 2002) but unsubstantiated to date. If syntax and words emerged simultaneously, no one has even tried to explain how or why they did this. From an evolutionary perspective, it seems obvious that words came first but had only a small subset of the properties of modern words, that their coming precipitated syntax, and that their subsequent interactions with syntax built the set of modern properties that Piatelli-Palmerini correctly lists but misinterprets.

Another protolanguage skeptic is Botha (2012), who finds the account of "living fossils" of protolanguage in Bickerton (1990b)—early-stage pidgins, utterances of trained apes, early child language, and the speech of "wild children" like Genie—less than convincing. About one of these categories at least (early child language) he is right, as shown in Chapter 7. But his arguments are beside the point, since the main case for protolanguage has always rested not on the existence of "living fossils" but on the logical inescapability of such a medium. As pointed out in Bickerton (1990b: Chapter 6), words must have preceded syntax, words had to be concatenated to make meaningful propositions, and without syntax they could not be merged pairwise (how would one know what to pair?) but could only be attached serially like beads on a string. Protolanguage could consist only of short utterances (no longer than three to five units, usually) *not* because there was some fixed upper limit to the process (a canard repeated in Chomsky 2005, 2010; Berwick and Chomsky 2011, despite the fact that no one ever claimed this) but because as nonsyntactic utterances lengthened, possible ambiguities multiplied exponentially.

Despite the fact that (for reasons still controversial) they lost out to humans, Neanderthals, our most closely related species, had brains on average slightly larger than those of humans (Stringer and Gamble 1993). It is tempting to hypothesize that this increase occurred because Neanderthals, still lacking the capacity for fully syntacticized language, required additional brain cells to compute all the possible meanings generated by attempts to create longer protolanguage utterances. That is, of course, sheer speculation, but it accords with the overall pattern of cognitive and cultural development in *Homo:* the relative cultural stagnation of the period between 2.5 and 0.3 million years ago and the commencement of a cascade of technological innovations shortly thereafter. It is consistent also with the need, after the first emergence of displacement, for a long period of brain restructuring before syntacticized language could emerge. This suggests a period of between one and two million years during which protolanguage gradually increased its vocabulary and developed from an enhanced communicative repertoire initially used only in confrontational-scavenging contexts to a fully symbolic system, but without undergoing other significant changes.

The issue of whether protolanguage was compositional, with units roughly equivalent to modern words, as assumed by Bickerton (1990b), Pinker (1994), Jackendoff (1999, 2002), Tallerman (2007), and others, or holophrastic, with units roughly equivalent to modern clauses, as proposed by Wray (1998, 2000), Fitch (2000), Arbib (2005), and others, has been discussed at length elsewhere (Bickerton 2009: 65–70; Arbib and Bickerton 2010). Here we may simply assume that the compositional account is correct, since the dissolution of holophrases into individual words presupposes an understanding of symbolism—that an arbitrary chunk of a holophrase could, in principle, represent the concept of some entity or event. There is no evidence for this, and, as we have seen, there is every indication that, prior to language, no animal had any notion even of the possibility of symbolic reference.

After Protolanguage

Although the infrastructure for language must have taken time to build, the transition from protolanguage to language could have been relatively rapid, taking place most probably as part of the transition from whatever species immediately preceded humans to anatomically modern humans. It is legitimate to ask whether a longer and more gradual shift would have been possible. There is no clear answer, but probabilities seem to go against the latter possibility. No condition exists intermediate between concatenating words like beads on a string and concatenating them by progressive pairwise merger. Until the full mechanism was up and running, message interpretation was likely safer by the "old" method (whereby words were dispatched individually to the organs of speech) than by the "new" method (whereby phrases and clauses were preassembled in the brain). The "new" method, if it worked well, was automatic and much faster than the "old" one. But if it didn't, garbled messages could easily result.

It is proposed here that the syntactic infrastructure resulted from self-organizing activity within the brain itself. Far from being a static structure, "the true nature of the cortex is dynamic, both within an individual's lifetime, and *within a species over time*" (Krubitzer 1995: 414, emphasis added; see also Kaas 1987). To illustrate the kind of course this dynamism takes, consider three species, one of which (the star-nosed mole) has a hyperdeveloped sense of smell that is shared by many other mammals, and two of which (the ghost bat and the platypus) have sensory modalities—echolocation and electroreception, respectively—shared by relatively few (Krubitzer 1995: fig. 2). Within an overall mammalian architecture that has remained essentially unchanged since the earliest mammals, the areas devoted to these senses differ dramatically in size and shape. Yet the dynamic self-organization of the brain, as Krubitzer points out, is such that "which features are likely to be retained, the types of modification that are likely to occur, and what will not happen can be predicted with some certainty" (416).

If we took the position that radical change originates within the brain by fortunate mutation and only subsequently impacts other organs, we would be forced to claim that the star-nosed mole first (for no particular reason) developed a capacity to process a wide variety of smells and only subsequently developed its olfactory sense; that the ghost bat similarly developed a mental capacity for echolocation, with the concomitant capacity to produce the requisite sounds not emerging until later; and that the platypus acquired the ability to process ambient electrical signals before developing electroreceptors in its snout and venturing into murky waters. Few if any biologists would accept this view. Most if not all would assume as a matter of course that new sensory organs began to emerge first, responding to forced or voluntary changes in behavior or habitat and that the brain struggled to play catch-up, adjusting itself so as to be able to process new types of incoming information.

In other words, a general assumption would be that when a new source of information becomes available, brains inevitably redistribute their resources, a process they can hardly accomplish overnight. Such changes do not need to be triggered by natural selection, although they may very likely result in improved reproductive capacity for those individuals that undergo them. They could be motivated simply by the brain's own requirements, such as shortening neural connections to save both time and energy. In the case of human ancestors, the new information was due not to the emergence of new sensory equipment but to the acquisition of words and their associated meanings. A dualist might assume that the "mental" nature of this new input would have to be treated differently from the "physical" nature of sensory inputs. This distinction is unlikely to have been meaningful to the brain. The brain translates every kind of input into the same language: a language of electrochemical discharges, their transmission, augmentation, and inhibition.

If there is any kind of innate, Universal Grammar in the human brain, it can consist only of structures and processes enshrined in the architecture of that brain. If we wish to trace the further evolution of language, and in particular those core elements that remain stable through all variability and change, the brain's most probable reactions to its colonization by words is the area we need to examine. If there are universals of language and cognition, it is only within the brain that they could have been given birth.

CHAPTER 5

Universal Grammar

"Universal grammar is dead" (Tomasello 2009: 470). This confident assessment came from a commentator on Evans and Levinson (2009a), a paper hailed by others too as being fatal to any innate universals specifically concerned with language. However, Tomasello is quick to note that "it is not the idea of universals of language that is dead, but rather, it is the idea that there is a biological adaptation with specific linguistic content that is dead" (471). Linguists have been looking for universals in all the wrong places. "Instead of looking at the input-output system . . . or the pragmatics of communicative exchange, they've been focused on the syntax and combinatorics, the least determined part of the system, as demonstrated by linguistic typology" (Evans and Levinson 2009b: 477).

Unfortunately "the syntax and combinatorics" are where the rubber meets the road. Without them, language would not exist. At best we would have remained stuck forever in the protolinguistic stage, where we left our ancestors at the end of Chapter 4. If subsequent events had really occurred without any biological support, we would have to face quite a different kind of evolutionary anomaly. We would be the first species that acquired its most crucial capacity without any biological adaptation to support that capacity. In other words, the anti-universalist attitude toward language that seems to have been growing and spreading in recent years is simply an anti-evolutionary attitude.

Nobody denies that there are features of the human vocal organs that have been biologically adapted for speech, and speech is no more than one of the output channels for language. If there were no comparable adaptations for what speech expressed, this would constitute a mystery that would surely baffle biologists. By rights, a dedicated adaptation should be the null hypothesis, and the burden of proof should lie on those who would challenge it. However, perhaps because of the specter of dualism, a dual standard seems to apply here. While philosophical dualists are hard to find nowadays, lingering dualist assumptions are harder to get rid of. Many, perhaps unconsciously, seem to assume some kind of difference between "mental" and "physical" phenomena. Vocal organs are "physical" and therefore legitimate targets for biological adaptation, but the neural areas and connections that subserve the structuring of sentences are "mental" and therefore somehow beyond the reach of biology. In reality the two sets of phenomena belong to a single class (call it "physical" or whatever you want to call it, it makes no difference), and they respond to biological factors in identical ways.

Evans and Levinson (2009a, 2009b) deal with a number of issues that allegedly challenge universals. For reasons of space I can refer here to only one of these: constituency. However, it is a crucial (arguably the most crucial) issue, and their treatment of it is typical of their articles as a whole.

Constituent Structure and Its Implications

Major claims of generative grammar are based on the proposition that language is formed on a universal basis of constituency. Constituents belonging to the same phrase must form part of a contiguous structure:

1. [A [friend [of [Bill's mother]]]] came here yesterday.
2. *[A friend] came here [of Bill's] yesterday [mother].

As indicated by the asterisk, (2) is totally ungrammatical in English, as it would be in any other language that obeyed the principle of constituent structure in its most literal sense. Similarly while clauses may appear in complex sentences with varying order, constituents of different clauses cannot be randomly interspersed:

3. [Although it was smaller], [the cat [that ate the rat] attacked the dog].
4. [The dog was attacked by the cat [that ate the rat]], [although it was smaller].

5. [[The cat [that ate the rat], [although it was smaller]], attacked the dog].
6. *[The cat attacked [although [that ate] it was smaller] [the dog] [the rat].

In (3)–(5) complete clauses (e.g., *that ate the rat*) can be inserted, as wholes, inside other clauses, but only if they serve to modify nouns in those clauses. With that exception, clauses must maintain their structural integrity. Sentences like (6) are ungrammatical in English and most perhaps all other languages.

However, Evans and Levinson (2009a: section 5) cite not only a number of "free word order" languages in which sentences comparable to (2) are fully grammatical but also, they claim, at least one (the Australian language Jiwarli) in which the same is true of sentences comparable to (6), in which constituents of different clauses can be jumbled. This evidence, they claim, destroys the credibility of any theory that takes constituency to be a universal of language. In what follows, I will assume the correctness of the facts they state, even though the authors produce neither an example of a real Jiwarli sentence like (6) nor a specific citation of where such sentences may be found, and even though the leading expert on Jiwarli mentions only intraclausal free word order in an article that deals explicitly with Jiwarli word order and case marking (Austin 2001).

If constituency were to be interpreted as meaning that all items related by constituency must *at all times* be physically adjacent to one another, then the facts Evans and Levinson adduce would indeed be problematic for generativists. However, the standard generative position has always been that constituency holds at some level of structure but that subsequent processes may produce variant surface orders. Note that the abandonment of the original deep-structure/surface-structure distinction (Chomsky 1957, 1965) by the Minimalist Program (Chomsky 1995) makes no difference as long as one assumes, as in the bare-syntax model of Chomsky (1994), that sentences are derived from the bottom up by a series of merges. Changes in the ordering of constituents can then be brought about in the course of the derivation by remerging a copy of the constituent to be moved and then deleting the original (e.g., if X is to be "moved," then [X] + [YX] → [X [YX]] → [X [Y –]]).

If we take the view that constituency has a conceptual basis that remains unaffected by surface considerations, it is the Evans and Levinson position that becomes problematic. The strongest evidence they adduce for the nonuniversality of constituency comes from Jiwarli. Recall their claim that Jiwarli can freely mingle words from different clauses. But the authors are forced to admit that the mingled words are "tagged, as it were, with instructions like 'I am object of the subordinate clause verb,' or 'I am a possessive modifier of an object of a main clause verb.' By fishing out these distinct cases, a hearer can discern the structure of a two-clause sentence . . . without needing to attend to the order in which words occur" (Evans and Levinson 2009b: 441). With these words the authors effectively nullify their claim. If constituents cannot be mingled unless the receiver can accurately and completely reconstruct the full original constituency of the sentence, then constituent structure is as real to that receiver as anything else in language.

Curiously, while showing that many previously claimed universals are not really universal, Evans and Levinson avoid mention of two universals that form part of the fundamental structure of every sentence in every human language. This is perhaps because nobody I know of has ever claimed them as universals, which must seem a surprising fact for nonlinguists (as it was for me when I first realized it). Perhaps it is precisely because these universals are so surface, so obvious, and so much taken for granted that linguists simply haven't thought of them or haven't thought them worthy of mention. Yet all sentences in all languages consist of phrases and clauses, and all phrases and clauses are constructed in essentially the same way, bear the same relations with one another, and enter into combinations of the same type. It is far from clear that this was inevitable. No known laws of nature enforce it, and at least one possible alternative—a language consisting solely of noun phrases—has been proposed and defended by Carstairs-McCarthy (1999).

Given an automatic mechanism for creating phrases and clauses, much of the rest of syntax follows logically or can be inferred, and what is left can be learned by good old general-purpose learning mechanisms. Some later sections of this chapter return to this issue, which will form an important theme recurring throughout the remaining chapters, especially Chapter 6.

The "Impossibility" of Universal Grammar

Some scholars go further than denying the existence of language universals. They claim that it is logically impossible for language universals to exist. The most comprehensive attempt to do this comes from Christiansen and Chater (2008). The position they start from is a strange one. Though the title of their article is "Language as Shaped by the Brain"—which, taken literally, is exactly the theme of this chapter—they mean by it something completely different. They propose that "language has adapted through gradual processes of cultural evolution to be easy to learn to produce and understand" (490).

Commentators on this article (among others, Harnad 2008; Piattelli-Palmarini et al. 2008) have pointed out some of the problems with this formulation. For instance, as it is stated it bears a "glass half-full/half-empty" relationship to orthodox Chomskyan proposals (surely counter to anything the authors intended), and it illegitimately reifies language as an extrahuman object with an identity, goals, and purposes of its own. More critically it logically entails that earlier versions of language must have been *harder to learn to produce and understand* than present-day versions! It follows that the forms of language most difficult to acquire must have been those that existed at its very inception. But how could it have gotten that way? Unless it was the gift of some benevolent deity or the manifestation of some Platonic ideal, language can only have come from the workings of the human brain. But if the brain originally produced it, how and why would language have had to make itself easier for that brain to process?

Ignoring these mysteries, Christiansen and Chater (2008) argue against the possibility of language universals. However, their arguments are based on tacit but highly unlikely assumptions: that universals began to be laid down when language was already fully functional and that this process lasted until relatively recently.

For instance, they argue that that because genetic change is far slower than linguistic change, grammars absorbed into the genetic code would be outdated before they could be put into operation. This assumes that the sole source for linguistic universals was prior linguistic production—that speakers first had to use certain types of structure and that these were subsequently encoded in the genome through some kind of "Baldwin effect" (Baldwin 1896). If that were so, they would be right, because linguistic production does constitute a moving target that no genetic process would be fast enough to track. But this argument cannot even in principle disprove the existence of UG. At best, it could merely show that UG cannot be produced by the Baldwin effect. But in fact things are much worse than that.

The very notion that UG could be produced by anything approaching a Baldwin effect (i.e., by incorporating genetically unsupported behavior into the genome) is itself incoherent. Any such process would entail that linguistic structures containing all the regularities to be included in the future UG already existed before that UG even appeared. If those structures did not come from that UG, where could they have come from? They could only have come from the human brain. What, then, would UG be doing that prior brain processes had not already done? Why would UG be there at all? Or, to put it more clearly, why couldn't you simply label as UG whatever enabled those processes? In which case, UG must have preexisted the structures that supposedly gave rise to it.

Other Christiansen and Chater (2008) arguments also fail because of faulty premises. For instance, they claim that because there has been relatively little contact between humans in different areas since the start of the human diaspora from Africa 60,000 to 90,000 years ago, we would expect to find "multiple UGs," different and conflicting versions of UG that would give rise to incompatible types of linguistic structure in the languages of different continents. But the absence of such versions depends on the unnecessary—and unlikely—assumption that the development of UG was incomplete at the time of the diaspora. If we assume that language has a long protolinguistic history and that (as niche construction theory suggests) significant behavioral changes are associated with speciation events, then the development of UG preceded or accompanied the last such event in the hominid line: the event that produced anatomically modern humans. Under such conditions, a complete UG would have preexisted the diaspora by tens of thousands of years.

Along with most others in the language-evolution field, Christiansen and Chater (2008) pay no attention to what the brain must have started doing when words were first invented. The only scholars who have focused on brain-internal developments are generativists (Piattelli-

Palmerini 1989, 2010; Jenkins 2000; Chomsky 2010). But there are profound differences between their approach to brain-internal developments and mine, which are now set forth.

The Brain Lends a Hand

Lest I be accused of rank inconsistency (since I have been so dismissive of approaches that see language as starting from brain-internal developments), I need to explain what those differences are and to demonstrate that there is nothing in the least inconsistent in saying that, while the process that led to language could have begun only in behavior, UG could have been derived only from the brain.

In generative theory, either language and UG emerged simultaneously, with no triggering external event, or UG preceded spoken language as a language of thought. In the present theory, UG and the enhanced communication that would grow into protolanguage emerged separately, in the reverse order. In other words, protolanguage emerged because of triggering external events: confrontational scavenging led to the need for recruitment, which in turn necessitated displaced communication, which eventually sufficed, in social animals with large brains, to create a crude and structureless protolanguage—all that nature needed. However, these processes necessarily caused symbolic items to be stored in the brain, and the only brain-internal processes we will be concerned with here are those that, while they may have exploited resources the brain already had, were directly initiated by the brain's need to deal with such items.

Note that this theory is fully congruent with the theory of niche construction (Odling-Smee et al. 2003). The process of niche construction may or may not be precipitated by environmental change, but it always involves behavioral changes. Nobody is claiming that these behavioral changes are what is uploaded into the genome, à la Christiansen and Chater's (2008) interpretation of Baldwin effects. What happens is that the consequences of the behavioral changes bias the selective pressures on the organisms concerned in such a way that those pressures now favor selection of any genetic variant yielding behavior that enhances exploitation of the new niche. In the present case, a change in subsistence-seeking behavior led to a change in communicative behavior that caused a significant change in selective pressures. These were pressures that affected the architecture and mode of functioning of the brain, and the resultant changes in brain structures and connections were subsequently incorporated into the genome.

The consequences of that incorporation turned out to be more far-reaching than those of perhaps any other evolutionary event since the emergence of multicellular organisms. Storage of symbolic units in brains made possible an orders-of-magnitude change in thinking processes, and the rest is, quite literally, history.

From the Start of Protolanguage

In protolanguage nature already had all it needed for a terrestrial, confrontational-scavenging ape to survive and prosper even under the most extreme savanna conditions. If the brain had not intervened at this point, probably nothing further would have happened. There would have been little if any motivation for language to become more complex. It is unlikely that any move in this direction would have increased the reproductive potential of the individual(s) concerned (Lightfoot 1991b; but see Progovac 2009 for a slightly different take on this issue). *Homo erectus* would have persisted without substantial change, finally—and perhaps only when environmental conditions changed radically—giving rise to descendent species with no substantially greater language capacity and/or going extinct. There would be no mysteries, no anomalies, no Wallace's problem. But things did happen in the brain, otherwise organisms like us might never have come into existence.

In most if not all work on human evolution, writers on both sides of the nativist-empiricist divide assume an essentially passive brain. For both Christiansen and Chater (2008) and Chomsky (2010), brain changes are random events. In Christiansen and Chater's case, they don't give rise to UG, and in Chomsky's they do, but in neither case do such brain changes directly respond to outside changes, nor do brains perform any action that might suggest they were driven by their own agenda. But there is good reason to believe that brain changes are anything but random and that a particular agenda does drive them.

Ever since brains began, they have had to adapt to new kinds of input. They have been adapting for thousands of millennia in order to more efficiently process data from a variety of sensory modalities, several of which are not shared by humans. Now they had to process data from a different kind of source, one that lay not outside but within themselves. This involved previously unrequired cortico-cortical connections, many of which had to be two-way, as well as considerable strengthening of existing connections between relatively distant parts of the brain. For example, a recent comparative study has shown that the arcuate fasciculus, a major pathway linking the lateral temporal cortex with the frontal cortex and long associated with language (Geschwind 1970; Catani and Mesulam 2008), is far more fully developed in humans than in macaques or even chimpanzees (Rilling et al. 2008).

The amount of work this entailed—for example, simply adding efferent to afferent fibers, and vice versa, since in many cases language data had to flow in two directions—should never be underestimated. It helps to account for, among other things, the length of time it took for protolanguage to become true language. But there is good biological evidence that the nature of the tasks that were required lie well within the brain's powers of self-organization.

A Brain-Driven Model of Syntax Evolution

The brain is notoriously plastic. That term is used here not (as it is often used in the neurological literature; e.g., Robertson and Murre 1999; Johansson 2000) merely to indicate the ability of the brain to reconstruct functions in new areas after original areas are compromised by trauma (although this ability is a corollary of what I am talking about). Rather I use the term in its lay sense to indicate a degree of flexibility and adaptability in both ontogeny and phylogeny greater than that found in other organs. What underlies and enables that plasticity is a relationship seldom discussed in the literature: the ratio of number of genes to number of neurons.

In a simple organism like the sea slug *Aplysia*, the ratio of genes to neurons is roughly 2:1 (18,000–20,000 to 10,000). In humans the ratio is of the order $1:4 \times 10^6$ (30,000 to 100 billion; Venter et al. 2001). In other words, while numbers of genes have varied relatively little across a broad phyletic range, increasing complexity has driven a massive increase in number of neurons. What this means in practice is that while, in *Aplysia,* it lies within the power of genes to determine the location and function of each neuron (just as they determine the location and function of limbs in all organisms), that power was soon exceeded by the proliferation of (mainly) cortical neurons. It follows that epigenetic factors rather than genes themselves mold the micro- and much of the macrostructure of the brain. One important factor is wiring economy (Cherniak 1994, 2005). To increase its efficiency while lowering energy expenditure the brain must reduce the total length of its wiring to the minimum necessary.

This does not mean that brain development falls outside of natural selection. What it does mean is that as long as there is variation within the brain (as there always is), evolution will select for what makes a fitter brain according to the brain's own needs, regardless of whether any selective pressure on the organism as a whole points in a similar direction. In plain language, while (as Lightfoot 1991b notes) a linguistic principle is unlikely to give Jack more fruitful sex, better organization of linguistic material and its processing would make Jack's brain work a lot better, thereby enhancing his overall fitness.

That the brain self-organizes in ontogeny is a well-known and well-established fact (e.g., Kandel et al. 2000: ch. 55; Nazzi et al. 2001; Mtui and Gruener 2006). What is less widely studied is the mode of development of the brain in phylogeny. As noted at the end of Chapter 4, while the overall architecture of the mammalian brain is strongly conserved, functions of the various areas are highly variable, and the variation is largely determined by which sense(s) predominate(s) in any given species (Krubitzer 1995). Thus when organs for hearing, say, improve in range and accuracy, brain areas devoted to processing sounds will multiply and expand, sometimes taking over areas formerly devoted to other senses, sometimes sharing auditory functions with the pre-existing functions of other regions (Krubitzer 1995; Manley and Clack 2004).

However, while brains can develop diversity within the bounds of a common architecture, they can just as easily achieve uniformity within that diversity. For example, Kaschube et al. (2010) studied the orientation columns (vertical arrays of neurons that preferentially respond to the same orientation) in the visual cortex of three species: ferret, tree shrew, and galago. The three species have been separated since the mammalian radiation of 65 million years ago and accordingly show wide differences in the size, shape, and location of the visual cortex and its relationship to other specialized areas. Yet at the same time the columns themselves are organized in almost identical ways in all three species. The authors attribute this too to the brain's capacity to self-organize. Indeed whether the brain shows diversity within uniformity or uniformity within diversity seems to depend not on any kind of external pressure but solely on the brain's ability to optimize its own resources.

Accordingly we can assume that as it began to store words, the brain underwent extensive reorganization. The consequences of that reorganization remain easily visible in modern human brains. Words are not stored randomly, but different word classes are stored in different areas; for instance, retrieval tasks suggest that nouns are stored in the temporal lobe, while verbs are stored in prefrontal areas (Caramazza and Hillis 1991; Damasio and Tranel 1993). This partitioning reflects, and presumably derives from, the conceptual difference between entities on the one hand and actions, states, and events on the other. However, this does not mean that their storage is confined to these areas. Rather the neural networks that subserve representation of a word include long- as well as short-distance cortico-cortical connections (Pulvermuller 2001). Thus networks supporting verbs like *kick* and *bite* will include neurons in the motor areas that control leg and mouth, respectively (Pulvermuller 2008).

But, the reader may well object, aren't I mixing up phylogeny and ontogeny? Surely while the overall redistribution of brain functions may be a phylogenetic process, the establishment of word networks can form only part of ontogeny. This is true, of course; we can hardly be born knowing the words we are destined to speak. However, since both the areas devoted to verbs and the areas devoted to leg control were established phylogenetically, and since the concept "kick" links to both properties, when the child learns *kick* its representation inevitably gravitates to those areas. However, the precise location of particular individual representations within the areas concerned varies unpredictably from one person to the next (Ojemann 1991). This follows logically from the fact that while overall areas for functions like "word class" or "similarity of meaning" are laid down in phylogeny, precise locations within those areas are determined by the individual child's particular course of development. (Varying orders in which words are acquired is one likely factor here.)

However, the most crucial contribution of brain developments to language lay not in the sphere of lexical parcellation but rather in that of perfecting and automating the construction of meaningful propositions.

From String to Sentence

One of the interesting results of attempts to teach language to apes is that in all species studied—chimpanzee (Gardner and Gardner 1969; Rumbaugh and Gill 1976; Terrace 1979), gorilla (Patterson 1978) orangutan (Miles 1994), and bonobo (Savage-Rumbaugh and Fields 2000)—the experimental subjects produced propositional utterances consisting of two or three units. As far as can be determined from the literature, the linkage of lexical items to form such utterances was never explicitly trained and though inevitably modeled in one sense (experimenters could hardly avoid using propositions in training) was neither modeled as an integral part of the training nor systematically reinforced in any way. In other words, these utterances appear to have been spontaneous. Some were probably imitations of previous utterances by trainers (as suggested in Terrace et al. 1979), but many were not, since they ordered units in ways that trainers never did. This suggests that the potential capacity to concatenate meaningful units must have been latent, not merely in the last common ancestor of humans and chimpanzees but in the last common ancestor of all pongids.

In any case, there are purely logical considerations that, once anything even approximating a word is used, will enforce concatenated utterances. Displacement signals differ from other forms of signaling (and resemble words) in that they relate to specific chunks of the environment rather than to desires, states of mind, or manipulative injunctions. Both words and displacement signals must link with other similar units if they are to function as transmitters of information.

Other kinds of signals can be used individually; indeed as individuals they are more useful than isolated symbolic units. If both you and I, reader, were vervets and I gave you the leopard alarm, you would know to run up a tree. But we are human, and if I say simply "Leopard," you can have no idea what I'm getting at. Since there are no leopards around, I could be starting (in a rather peculiar way) to tell you about some personal experience I had with a leopard, or I could be about to give you some information I had learned about leopards, or I could be quoting a familiar essentialist adage—though in the last case, you might wonder why I omitted the indefinite article in "A leopard can't change its spots."

Consequently the development of simple concatenation in protolanguage requires no special explanation. What does require explanation is how words came to be concatenated in a special way and subject to language-specific principles—in other words, how sentences were created.

Concatenating Symbolic Units

Establishing locations in the brain was not the only task that words obliged brains to undertake. The owners of those brains seemed to want to assemble those words, but at first they could do this only in an ad hoc, structureless manner—the mode of protolanguage. Doing things that way wastes energy, and brains place a high priority on energy conservation. "A universal feature of human behavior is that well practiced tasks can be performed with relatively little effort or cognitive control, whereas novice performance of the same task may require intense and effortful cognitive control" (Poldrack et al. 2005). Consequently brains must have sought to automatize the process of utterance. And automatization consists of fixing on a stereotyped routine and then increasing the rapidity with which that routine can be executed.

But how was this to be done? With regard to the most fundamental process of the routine—linking a single pair of symbolic units—it is seldom taken into account that there are two significantly different ways in which the brain could do this. The groups of neurons that represent words could directly trigger one another, so that representations of two or more words could be linked in the brain before being dispatched *as a sequence* to output channel (oral or manual) for utterance. Alternatively the representation of each word could be retrieved separately and sent *individually* to an output channel for utterance before the next word was similarly retrieved and dispatched. Nothing in the structure of the resultant sequences would tell a receiver which method had been used.

However, we can predict that the first method, involving only a single (albeit more complex) set of instructions to the output channel, would result in observably faster speech than the second method.

It is clear that language uses the first method. Recall the examples in the postscript to Chapter 2, repeated here for convenience:

7. Who do you want to meet?
8. Who do you want Mary to meet?
9. Who do you want to meet Mary?

In casual speech, the sequence *want to* can reduce to *wanna* in (7):

10. Who do you wanna meet?

Obviously *want-to* contraction is blocked in examples like (8), where *want* and *to* are not contiguous. But it is also blocked in (9), where they are contiguous. Why? Because in (9), *who* has been moved from the position occupied by *Mary* in (8):

11. Who do you want [who] to meet Mary?

(Compare the surprise/confirmation-request question *You want who to meet Mary?*) Suppose the brain was selecting and sending words to the speech organs one word at a time. In (7)–(9) that fourth word would be *want*. Unless the sentence was preplanned, the part of the brain that directly controls speech organs would have no way of knowing what the fifth word would be. It could be *on* ("... on your team?") or *if* ("... if Mary can't come?") or *for* ("... for a partner"). If the following word is anything other than *to, want-to* contraction cannot apply. But if the following word is an unpronounced copy of the subordinate-clause subject, it cannot apply either. The brain's computational mechanism must somehow know already that a subordinate clause must follow the fourth word and that the subject of the clause, being a questioned element, must have left an unpronounced copy between *want* and *to*. This entails that the whole sentence must have already been constructed before being spoken.

Some empirical evidence comes from the differences in utterance time between trained apes and humans, and between pidgin speakers and speakers of a full human language (who may, of course, be the same person, at different times). Typically, long intervals intervene between items in ape discourse; thus, for example, Washoe's celebrated utterance *baby in my drink* was spread over three seconds (Terrace et al. 1979: table 7), whereas three words per second is a normal speed for human utterance. Similarly in early-stage pidgin, speed of speech is about three times slower than the creole speech of pidgin speakers' descendants (Bickerton 2008; the relevance of this will become apparent in Chapter 8).

These speed-of-utterance differences surely reflect different forms of neural connectivity. (It is not easy to see what else they could reflect.) Slow speed results when either there is no viable connection between representations of words or whatever connection exists is so long and roundabout that it is quicker and more efficient to pronounce words serially and individually than to assemble them before dispatch. Direct, fast connections between representations of words in the brain are formed by repeated simultaneous firing of the relevant neurons (Hebb 1949; see also Pulvermuller 1999; Garagnani et al. 2009 and references therein).

It therefore follows that members of several populations cannot have direct, fast connections. These populations include persons in the earliest stages of second-language learning, early-stage pidgin speakers, and human ancestors at the start of language evolution. In none of these cases have subjects had time to accumulate the number of simultaneous firings required before direct linkages can be formed. In human ancestors that stage was never reached. In pidgin speech it can be reached in principle but relatively seldom is. In second-language learners, whether or not it is reached will depend on the degrees of exposure and motivation of individual learners.

In other words, uttering words is an autocatalytic process: the more you use them, the quicker you become, to a point where you can mentally assemble utterances before producing them. This is probably what underlies the anecdotally well-supported but understudied phenomenon known as "thinking in a foreign language," a stage characteristically not attained until at least a moderate familiarity with the lexicon of the language involved is in place. A counterexample might seem to be what, from its most famous exponent (the infant Lord Macaulay), one might call the "Macaulay effect": another well-supported but even more understudied phenomenon in which a child remains mute until age three or so, then begins to use complete phrases and clauses. But for this to happen, others have to have been speaking, and the child has to have been processing their language. Apparently passive processing is enough to promote simultaneous firing of lexical representations (Petersen et al. 1990; Price et al. 1996), thus enabling the formation of direct links between them that enables rapid utterance.

Linking words to form propositions does two things. First, since words have to be linked to their associated concepts, the simple act of uttering them creates links that join concepts as well as words. This is precisely what enables the voluntary and rational trains of thought that distinguish human from nonhuman thinking. Second, the conjunction of typical pairs of words—for example, *run* and *tree*, a verb of motion plus its directional complement—lays the basis for syntax by prioritizing certain pairs of words and binding them more firmly than pairs that lack any strong semantic or pragmatic connections. Such conjunctions could have served as precursors of Merge, the basic process of syntax. (Recall, however, that some Merge-like process is implicit in how all brains handle vision, combining the outputs of a variety of cell assemblies to create an apparently seamless picture.)

Structuring Concatenations

In protolanguage there could have been no formal constraints on which words could be linked with which. It is likely too that protolinguistic utterances made frequent use of repetition. This is common in ape-language experiments; take, for instance, the incident in which Panbanisha "repeatedly pressed three symbols—'Fight,' 'Mad,' 'Austin'—in various combinations" to report a fight between two other apes (Johnson 1995), or the longest recorded ape utterance, "Give orange me give eat orange me eat orange give me eat orange give me you" by Nim (Terrace 1979: 210). In a similar vein, McDaniel (2005: 162) hypothesizes *baby tree leopard baby baby kill* as a possible protolanguage utterance by someone observing a leopard about to leap from a tree upon a baby beneath. As a warning this might work, but for everyday communication a little pruning would help. Given repetition and no syntax, the ambiguities of utterances quickly multiply. It would be difficult, perhaps sometimes impossible, to determine who did what to whom.

As soon as one gets past warnings and commands in any kind of language, it becomes essential for comprehension to determine who did what to whom. But without syntax, one has to depend on pragmatic or semantic clues, which may be sparse or absent or, if present, confusing. It would not be surprising if protolinguistic utterances took on some of the rough consistencies of word order found in pidgins (Givon 1979), in which topics and old information typically precede comments and new information. Such overall consistencies have been found in ape utterances (see Savage-Rumbaugh and Rumbaugh 1993 for a review), with numerous exceptions, naturally. This factor may have helped comprehension but could have been useful only in utterances that would correspond (roughly) to single-clause sentences. Anything longer would require a stereotyped, repeatable, automatic structure in order to ensure accurate comprehension.

To what extent is it realistic to suppose that a brain whose only goal could have been to maximize its own processing efficiency could have provided structure to make language more efficient for prehumans? One interesting suggestion comes from Hurford (2003, 2007b): many nonhuman organisms share with humans two brain pathways, dorsal and ventral, which respectively determine the location and the identity of objects, and Hurford sees this as the basis of predication—predication in turn forming the basis of syntax. However, "the origin of the terms subject and predicate belongs to logic, not grammar" (Jones 1956: 184), and structures with more than one participant (subject, direct and indirect object, etc.) form by far the most frequent type of clause. A commitment to gradualism might suggest that the brain began with simple predications like *lions (are) dangerous* and then slowly worked up to more complex structures like *looks like the lions round that dead dinotherium have already eaten the best bits of it*. However, there seems no reason why, given nouns and verbs, the brain should not have elaborated a systematic mechanism for automatically generating sentences without needing to build on some simpler precursor, even though (as was likely the case) initial products of that mechanism may have tended to be shorter in length.

Until the advent of brain-scanning techniques (for a recent historical overview, see Gulyas 2010), it was possible to believe that language was mediated by one or two highly specialized and localized modules. Early research on aphasia seemed to implicate particular loci (Broca's and Wernicke's areas), with Broca's area supposedly handing syntactic aspects of language while Wernicke's handled semantic aspects. Results of PET scans and fMRI, however, showed that if there was anything resembling a "language organ," it was more like a "language amoeba" (Szathmary 2001), a network of connections linking numerous and diverse brain areas, including extracortical regions. Indeed the network *specifically devoted to syntax* was almost as widespread: "The focal areas of the network (Broca's area, Wernicke's area, Cerebellum, DPF [dorsal prefrontal cortex], A[nterior] C[ingulate] C[ortex]) are very strongly interconnected" (Dogil et al. 2002: 85). None of the areas in question is exclusively devoted to syntax, and all are incorporated in other networks that discharge different functions, so it has been difficult to sort out the precise contribution made to syntactic processing by each of the areas concerned.

Work by Friederici and colleagues (Grodzinsky and Friederici 2006; Friederici 2010) has proposed two distinct structures, one dealing with local phrase structure, another with more complex sentences. However, most imaging studies concentrate on comprehension rather than production, and while comprehension can call on other aspects of language besides syntax, production invariably demands that syntactic processes be employed. Unsurprisingly almost everything about the brain's syntactic processing is either vague or controversial, and equally unsurprisingly a consortium of area specialists recently agreed that "to understand the neurobiology of syntax, it might be worthwhile to shift the balance from comprehension to syntactic encoding in language production" (Fedor et al. 2010: 309). To which the working syntactician can only add a fervent "Amen!"

An Abstract Model of How the Brain Handles Syntax

How might a brain have gone about the task of devising a formula for the construction of sentences? To start with, all the brain had was a small number of things it had already sorted into separate classes, one denoting entities (which would become nouns) and one denoting actions, states, or events (which would become verbs). It is not a natural necessity that things should have turned out this way. Carstairs-McCarthy

(1999) has envisaged a state of affairs in which there were no verbs, so that instead of sentences language would have consisted entirely of expanded noun phrases. Moreover he has shown that such a language would not necessarily have less expressive power than the kind of language we know.

What does seem natural is that, given the existence of two classes, each class should become the basis for structures of increasing size—that if there were nouns and verbs, there should eventually be larger units formed by adding other words to nouns and verbs, respectively. After all, almost any word needs an accompanying word or words if anything less that the broadest of generalities is intended.

Traditionally these larger units are described as "noun phrases" and "verb phrases." However, these expressions have meanings that are, frankly, just a confusing nuisance for the exposition ahead. First of all, they entered into the common assumption (Chomsky 1957) that "noun phrase" and "verb phrase" are distinct and separate complementary elements of propositions. Second, their meaning has undergone significant changes in the past few decades. What was originally called a noun phrase is now often called a "determiner phrase" (determiners being articles, demonstrative adjectives such as "that," and the like) in which a determiner takes a noun phrase as its complement (Abney 1987). Earlier, subjects of sentences were excluded from the verb phrase, but more recently they have been moved into it (Koopman and Sportiche 1985). Consequently I will abandon the terms *noun phrase* and *verb phrase* and refer to a noun and all its modifiers as simply "a phrase" (no other kind of phrase will be either discussed or assumed) and to a verb and all its modifiers (which of course include all its arguments—participants in the verb's action—whether "internal" or "external," obligatory or optional, including its subject, as well as adverbs and the like) as "a clause," unless others' uses of these terms are being cited.

The necessity for expansion to phrases in the case of nouns is due to the fact that a noun by itself cannot specify an individual (unless of course it is a proper name) but only the concept of a class of entities. As soon as protolanguage was divorced from the here and now—that is, when immediate context plus pragmatic considerations no longer sufficed to identify specific referents—modifiers became necessary: *that dog, the brown dog, the dog with the wiggly tail, the dog I told you about yesterday.* Modifiers could grow in length and complexity as the system grew precisely because there was no specification of what could constitute a modifier or (beyond purely practical considerations) how long a string of modifiers could be.

Verbs too needed modifiers. Naming a class of actions, states, or events tells us next to nothing. We need to know what participants were involved in the event, who or what experienced the state, who performed the action, and who or what was affected by it. To specify any occurrence we need to know not just the verb but the key participants in whatever occurrence the verb specifies, from one to three in the vast majority of cases (corresponding to what, in grammars, are called "subcategorized arguments" of the verb). Here at least two preexisting factors intervene: episodic memory and reciprocal altruism.

Episodic memory was originally supposed to be limited to humans (Tulving 1984). However, the present century has shown increasing support for the existence of nonhuman episodic memory from both theoretical (Dere et al. 2006; Griffiths et al. 1999) and empirical studies. Among the latter, what is cautiously termed "episodic-like" memory has been demonstrated in scrub jays (Clayton et al. 2001), rats (Eacott et al. 2005), and hummingbirds (Henderson et al. 2006), among other species. Episodic memory could supply the awareness of event structure ("who did what to whom") that forms a basic essential of syntactic structure.

That it already did this in at least some alingual species is suggested by the theory of reciprocal altruism (Trivers 1971): "I'll scratch your back if you scratch mine." Game theory (Maynard Smith 1982) entails that for reciprocity to become an evolutionarily stable strategy, there must be some mechanism for detecting cheaters (Barkow et al. 1992). Such a mechanism would of course be available if episodic memory did indeed provide event structure. Some biologists remain skeptical on this score (Hauser et al. 2009), but there is impressive evidence of reciprocity in grooming from Seyfarth and Cheney (1984) to Schino and Aureli (2008). The latter surveyed grooming reciprocity in twenty-two primate species and found that in all species "female primates groom preferentially those group mates that groom them most." (Hauser et al. 2009 dismiss these cases as susceptible to a more parsimonious explanation, although they do not say what such an explanation might consist of.)

Grooming preference among primates generally is thus most likely guided by episodic memory of both "how often you groomed me versus how often I groomed you" and "how often X groomed me as compared with how often Y groomed me." This is sufficient to yield event structure, "who did what to whom," which in turn provides the basic structure of clauses (Bickerton 2000).

The addition of modifiers to verb heads would thus require the output of the noun-phrase building operation already described. The semantics of each verb would then determine whether it would require one *(Mary left)*, two *(Mary met John)*, or three *(Mary gave John a book)* phrases. In addition, optional items might be added; even those still skeptical about full episodic memory in nonhumans seldom dispute that in many animals "episodic-like" memory supplies "when" and "where."

But the crucial question is this: Given these materials, how could the brain produce syntax out of the random, pragmatics-driven utterances of protolanguage? One likely way emerges when we consider how the brain deals with routinized motor actions such as aimed throwing.

I have to be careful here, because what I am saying could be misinterpreted in two ways. It could be taken as endorsing ideas expressed in the work of Lieberman (1984, 2010) and Allott (1992), which assume that motor systems directly underlie language mechanisms and were directly responsible for their evolution. Or (especially as I raised the paradigm case of throwing) it could be taken as endorsing proposals by Calvin (1982; Calvin and Bickerton 2000) that human cognitive powers derived in large part from aimed throwing and that a sentence is somehow structured like a throw. These approaches are, I think, overly simplistic, but there is a more general sense in which the structuring of a sentence is analogous to the structuring of a stereotyped motor action. Both must have a schema (Arbib 2003), a predetermined program that dictates (among other things) the sequence in which the component parts of actions must be executed for maximum effect. For instance, the launch window for a throw is only a few milliseconds long, but it must be timed so as to coincide with the highest point in the arc of the throwing arm.

Structuring and automation of motor actions must have a phylogenetic as well as an ontogenetic perspective. It cannot be the case that baby chimps and human infants start from an equal level of throwing skills, so some kind of schema for throwing (one that still, naturally, requires to be fine-tuned by experience) must form part of the human genome. The brains of human ancestors must have begun from a chimpanzee level of throwing. They could have developed from that level only if aimed throwing, whether used in confrontational scavenging, opportunistic hunting, or intergroup warfare, enhanced human fitness. All I am proposing here is that, once our ancestors started to talk at even a protolinguistic level, the brain would have begun to elaborate schemas for routinizing sentences (call them algorithms for sentence production, if you prefer), just as it elaborated schemas for throwing and similar actions. Stereotyping and automatization of action schemas save energy and free up neurons for other functions.

For some it may seem easier to believe in schemas for physical action than for what they regard as "mental actions." This is another reflex of persistent dualism. Whether we are talking about physical actions like throwing, "mental" actions like reasoning, or things that seem to blend both mental and physical, like creating and uttering sentences, we are talking about events that are triggered by exactly the same operations: sets of neurons firing other sets of neurons. It has been known for some time that the neurons fired when an action is performed are the same neurons fired when one is merely thinking about performing the action (Decety et al. 1991, 1994; Paus et al. 1993; Jeannerod 1997). In other words, as far as the brain is concerned, thoughts are just actions that it doesn't tell the body about. However, we should be cautious about going from this to the statement "Thoughts are just sentences that the brain doesn't send out for utterance." A thought will not have all the clutter of modern-language sentences—the inflections, the function words, the phonological features. But the means for linking the concepts that thought manipulates will be the same as the means for linking words into phrases and clauses that language manipulates. The syntax of thought will be the same skeletal UG that is being described here.

To sum up, we can assume that at a minimum, a primate brain, once equipped with words, would eventually create algorithms for the construction of phrases (take a noun and add modifiers) and clauses (take a verb and add phrases). Note that, among much else, these algorithms entail the infinite productivity that ever since Chomsky (1965) has been recognized as basic and essential to language. Note that this productivity is at least as much due to properties of the phrase as it is to properties of the clause. A clause has a built-in limit. It can include only as many participants as the brain can distinguish functions for (i.e., entities with thematic roles): Actor, Theme (what the action of the verb directly affects), and Recipient, the three key functions picked out by the episodic-memory/reciprocal-altruism devices, plus Time, Location, and one or two others. However, there is no limit to either the kind or number of modifiers a noun may have, including clauses *(the black dog with white spots that you saw yesterday)*. (Of course, verbs too can have clauses as modifiers if they fulfill one of the verb's required functions, e.g., Theme in the case of verbs that take factive complements: *you told me that you came yesterday*.) Since a verb that takes a factive complement can select a clause with a similar verb as its head *(John thought that Mary felt that Bill believed . . .)* and since a noun inside a modifier clause can itself be modified *(This is the cat that killed the rat that ate the malt that lay in the house . . .)*, sentences are, in principle, infinite.

All that remains is to show how these algorithms could work in practice to produce the kinds of linguistic phenomena that syntacticians have described.

The Process of Sentence Construction

Although Chomsky's Minimalism was never intended as a model of how the brain puts sentences together, it serves, as we saw in Chapter 2, as a good start toward understanding how this might work.

Whatever your theoretical assumptions, the brain must begin sentence construction with some equivalent of what Chomsky (1995: 225) calls the "numeration," the collecting of all the lexical items to be used in a sentence (in neural terms, the simultaneous firing of neurons in the areas where the words to be used are stored). It is followed by the operation "Select," which begins by choosing from the numeration the first items to be merged into the derivation.

How are elements for the numeration selected? The universal and well-attested phenomenon of "lexical priming" (Bock 1986) suggests that in the real world the brain may be unable to call on any lexical item without triggering others, some of which may be irrelevant to the sentence in question. It may be that the brain obeys some kind of "neural

Darwinism" (Edelman 1987) such that, as Calvin (1996) suggests, there may be several "candidate sentences" forming at the same time and competing for access to the speech organs. This is consistent with speech substitution errors due to similarities of sound, absent semantic similarity ("fire" for "flour," "slips" for "sleeps," "clam" for "climb"); semantic relationships, absent phonological similarity ("degree" for "job," "next" for "last," "science" for "medicine"); and even, where both phonology and semantics are similar, for blends such as "swishle," a combination of "swish" and "swizzle" (data from Garnham et al. 1981). The notion of multiple selection followed by pruning of some kind (i.e., initial selection is random and only items that match features with one another are selected) is plausible in terms of "neural Darwinism."

The operation that Chomsky has described as "Merge" then follows. For Chomsky (2000), the procedure "Merge A and B" is a symmetrical operation. Different, and I think preferable, consequences emerge if we regard the operation as asymmetrical, redefining it as "Attach A to B," A being a modifier and B a head (or a structure that already incorporates a head and one or more modifiers). Here B clearly has priority over A. Since any word functions in any given expression as either a head or a modifier (depending on function rather than word class: *city* is a head in *capital city* but a modifier in *city council*), "Attach A (a modifier) to B (a head)" covers all possible constructions. One advantage of this operation is that it does away with the need for any special mechanism for determining the label that its product will bear. If the head is a noun, then no amount of modification can change its nominal status, or the label of the structure it heads; if a verb, a similar conclusion follows. Two things are worth noting about Attach: like Merge, it does not specify direction of attachment (whether to the left or right of the head), and like Merge, it places no limitation on the type or sizes of units to be attached.

For each phrase, the brain would begin by selecting a noun and would then progressively attach to that noun whatever modifiers in the numeration applied to it. Similarly for each clause, the brain would select a verb and then take phrases modifying that verb (as they were produced by the phrase-forming processes of Attach) and progressively attach them to the verb. Note that (and this is important for what follows), counterintuitively, a derivation of this kind must proceed not in the linear order in which the words will eventually be uttered but from the bottom up, so to speak—from the lowest level in the hierarchical tree structure that will constitute the finished sentence. This is because words must initially attach to the things with which they are most closely associated: direct objects to their verbs, for instance (in English, if not in French). Note too that adjectives are attached in a determined order:

12. a clever young Spanish English teacher.
13. a clever young English Spanish teacher.

These phrases are not synonymous: (12) means a teacher of English who is Spanish and (13) a teacher of Spanish who is English. Other variants are simply ungrammatical in either sense:

14a. *a clever English young Spanish teacher.
  b. *a young Spanish English clever teacher.
  c. *an English clever Spanish young teacher.

Clearly order of attachment is a crucial factor in sentence creation. Moreover it is an ineliminable component of language—words have to be attached to one another in some particular order—but one that is not specified either by Merge or by Attach.

Both Merge and Attach demand some kind of checking process to ensure a semantically and grammatically acceptable combination of words. Checking basically consists of comparing features on the items to be attached to ensure that there is a match. A singular noun cannot accept a plural modifier (e.g., *several dog). A verb that takes only a single participant cannot accept phrases that represent Agent and Patient roles (e.g., *Mary fell Bill as opposed to Mary fell or Mary knocked Bill down). Similarly needs of a head have to be met by a modifier: the verb *see* (except for idiomatic usages like *yes, I see*) must have a direct object with a thematic role of Patient for whatever was seen. If any of these conditions is not met, an illegitimate sentence results.

Order of attachment is not written in stone, because "thing most closely associated" is subject to interpretation. Is a verb more closely associated with whatever the verb affects or with an adverb that modifies it? English takes the former position, French the latter. Or is it whatever the clause is about—the subject as opposed to the predicate? That is what Hawaiian chooses. This last difference, the difference between subject-verb-object and verb-subject-object order, is less important than one might think, for as shown below, the first and the last attachment to a clause have very similar consequences.

The foregoing, while incomplete in many ways (a full account would require at least book-length treatment) and possibly quite wrong in some points, at least represents an attempt to relate necessary processes in the construction of sentences to known capacities of the human brain. The question that naturally now arises is how much syntax we can buy with mechanisms such as those described above. Ideally all truly universal features of syntax should follow from the model described, and the grammar sketched here, while not generating any modern human language, would have yielded a viable language. Since it is universal, it serves nowadays as the common skeleton underlying all modern languages. It is what enables children to acquire any of the several thousand languages in the world today and, in appropriate circumstances, to build an entirely new language for themselves.

Consequently the sections that follow should not be read as any kind of attempt to establish a definitive, detailed grammar but rather as proof of concept, showing that the kinds of structure and operation proposed here are at least potentially able to deal with data that have hitherto required more complex assumptions and/or a greater amount of stipulation. Readers should, however, be warned that the rest of this chapter is somewhat technical. That is unavoidable. No serious syntactician gives any credence to programmatic statements, and rightly so. At least some indication of how the model would work in detail has to be provided at this point. Readers who find the technicalities tedious or overly challenging can skip to the concluding section of this chapter without undue loss. Meanwhile let's look at three areas—c-command, island effects, and the reference of "empty categories"—which should suffice to give at least a flavor of the present proposals.

C-command

Although c-command has proven a tough survivor among generative notions, it remains unmotivated. As Hornstein (2009) points out, Epstein's (1999) deduction of c-command from Merge makes the concept natural in Minimalist Program (MP) terms without explaining why it should necessarily work the way it does. (Hornstein's own explanation—that c-command is the kind of relation one would expect from a Merge-based Minimalist grammar—is hardly an improvement.) In the present model, the utility of c-command is dubious. It is not required for determining the order of constituents (Kayne 1994) because Kayne's Linear Correspondence Axiom itself is not required. It is not required as a condition on feature checking because this now takes place via projection of head features through higher structural levels of phrases. In binding, where it has long played a vital role, it has always encountered serious empirical problems.

In standard generative models, to bind an anaphoric expression like *herself* in *Barbara admired herself* the anaphor has to be c-commanded by its antecedent (as it is here by *Barbara*). The same goes for bound pronouns such as those associated with *each X* or *every X*. But these relations are asymmetric: where the anaphor c-commands its antecedent, ungrammaticality results:

15a. Barbara admired herself.
  b. *Herself admired Barbara.
16a. Each boy received his prize.
  b. *His prize was given to each boy (* if *his* and *each boy* are coreferential).

Note, however, that in these cases the c-commanding antecedent is a subject. When it is not, the following phenomena occur:

17a. Mike showed Barbara herself as she really was.
  b. *Mike showed herself Barbara as she really was.
18. I gave each boy his prize.
  b. *I gave his prize to each boy (same condition as 16b).

The most straightforward tree for the grammatical (a) sentences would merge first *showed/gave* with *Barbara/each boy* and then *showed Barbara/gave each boy* with *herself/his prize*. But in that case, the antecedent would nowhere c-command the anaphor. Indeed in (17a) and (18a) it would be c-commanded by the anaphor. Barss and Lasnik (1986) treated the problem as if it were limited to double-object constructions, but it holds equally between objects and adjunct phrases, witness (19):

136   **Universal Grammar**

19a. Bill gave the money to Barbara for herself (not for her sister).
   b. *Bill gave the money for herself to Barbara.

In neither case does *Barbara* c-command the anaphor, but in the grammatical (19a) *Barbara* is attached before *herself*, while the reverse holds in the ungrammatical (19b).

Larson's (1988) proposal to deal with this problem aimed specifically at double-object cases, but could be (and soon was) extended to other cases. As with so many other proposals, this one involved covert movement. Additional empty nodes are created; the verb moves to a higher V node and the NP to a higher NP node, giving the structure shown in Figure 5.1.

In this configuration, known as a "VP shell," c-command of I(ndirect) O(bject) by D(irect) O(bject) holds in the original structure, while subsequent movement of NP(IO) to NP and lower V to the higher V position, as shown, provides the required surface order. The (to me) irresistible question is this: If the mechanism expressed by this tree does not represent anything the brain is supposed to do in the course of forming such sentences, what is it doing in the grammar? Conversely, if the brain does indeed perform this movement operation, why should it? Just to salvage c-command?

Larson's (1988) solution, widely adopted, cannot form part of the present model, since the latter avoids any operation (such as covert move-

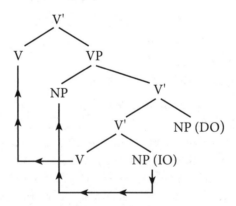

Figure 5.1. A VP-shell structure.

ment) for which empirical data provide no support. Covert movement has simply been accepted as part of the "biolinguistic" approach without, so far as I am aware, any discussion of its connection, if any, with how brains actually produce sentences, how grammatical operations might have evolved, or any other substantive issue that an approach aimed at unifying linguistics with biology might have been expected to pursue. Moreover there are other failures of c-command that Larson's solution cannot resolve. Consider (20), where a constituent containing the anaphor c-commands the antecedent and where no plausible and independently motivated covert movement will remedy the situation:

20. Stories about herself make Barbara feel deeply embarrassed.

However, an attachment-based model necessarily creates two relationships, both of which are based on the sequence of attachment and are therefore both natural and ineliminable. These relationships, priority and finality, between them discharge the functions formerly attributed to c-command, as well as others.

Priority arises because attachment must be sequential and therefore establishes an order among constituents that is as fixed and definite as the linear order in which sentences are spoken or written. It is in at least one sense a more significant order, since it is directly based on the degree of structural closeness constituents have to one another. Priority may be formally defined as follows:

21. X is prior to Y iff X is attached to a structure Z before Y is attached to Z.

Finality is significant because of another coincidence between the present proposals and one of those elaborated within the MP. As we saw in Chapter 2, Chomsky (1999) envisaged sentences as falling into distinct "phases" which are sent separately and sequentially to the speech organs and cannot subsequently take part in any syntactic process. Sentence components are assembled simultaneously and in parallel, and full assembly of a verb and all its modifiers (auxiliary verbs, arguments, and any adverbial material), just like completion of a Chomskyan phase, renders the resultant syntactic object computationally inaccessible. One argument must logically be final, that is, the last one to be attached to a particular verb. Finality may be formally defined as follows:

22. X is final in a structure Y iff X is referential and no further referential element can attach to Y.

Note that a final attachment asymmetrically c-commands all other constituents in its clause and thus accounts (among other things) for the cases in which c-command works correctly in binding (e.g., (15) and (16)) while excluding those cases in which it doesn't. In (17a) and (19a), *Barbara* and *each boy* are prior to *herself* and *his prize,* respectively, while in the bad (b) equivalents the relationship is reversed. In (19a) *Barbara* is prior to *herself,* while the reverse holds in (19b). In (20) the constituent that includes the anaphor c-commands the antecedent, forming a problem impossible to resolve unless such sentence types are arbitrarily stipulated as falling outside binding theory. However, it is straightforwardly resolved once we accept that priority as well as finality binds.

Note that finality and priority also take care of another case where c-command requires covert movement. Quantifier Raising has a long and respectable history in generative grammar, from May (1977) to Hornstein (1995). It arose in order to account for the identical interpretations of sentences such as (23a) and (23b):

23a. Every girl kissed a boy.
  b. A boy was kissed by every girl.

Both sentences can mean either "Every girl kissed one particular boy" or "Every girl kissed at least one individual boy or other." From the beginning, the issue was phrased as one of scope, and scope was supposedly a function of c-command. That is to say, if a phrase with a quantifier like *each boy* or *every girl* c-commanded an indefinite phrase such as *a boy* or *some girl,* that phrase could then be interpreted in a general as well as a specific individual sense. The converse, of course, should not apply, so that (23b) should have had only the first reading. Since it did not, the reasoning then went as follows: there must be some point in the derivation of (23b) where *every girl* c-commands *a boy.* In the earliest stages of generative grammar, where passives were transforms of actives, there was no problem: the quantified phrase c-commanded the indefinite in Deep Structure. But when it was decided that the passive transformation changed meaning and was therefore inadmissible, an alternative had to be found. Since meaning was involved, the logical answer was to invoke Logical Form (LF, the level at which sentences are interpreted) and move the quantified phrase into a c-commanding position in LF. This did not change substantially when LF became the C(onceptual)-I(ntentional) or SEM(antic) Interface, so that there was only one syntactic level. On that level, movement (overt or covert) took place, and if the movement was covert, the top copy was retained at the C-I interface, but dropped at the S(ensory)-M(otor) Interface, where the sentence would have to be pronounced.

As the above suggests, this approach had more to do with how generative grammar happened to develop than with the data to be explained, let alone the overarching problem of how syntax can be made congruent with things the brain does. Why should one configuration of quantifier/bound-item be treated one way and another in a different way? Why shouldn't the interpretations of (23a-b) fall out from the basic mechanisms of syntax? In orthodox varieties of generative grammar there are no answers for such questions. In the present approach, these sentences are simply further examples of general binding constraints. In (23a) *every girl* binds *a boy* by finality, *every girl* being the final attachment. In (23b) *every girl* binds a boy by priority, *every girl* being attached before *a boy* is attached.

One obvious objection to a priority/finality analysis is that it has to invoke two mechanisms, whereas orthodox accounts involve only one. But this objection runs counter to the spirit of the MP. It is not a reduction of mechanisms per se that should count in favor of a theory but whether the mechanisms are inevitable in assembling sentences, whether they account naturally for the phenomena they are claimed to account for, and whether their adoption makes it possible to dispense with previous proposals that are more complex and/or less natural. Both c-command and priority/finality are inevitable consequences of hierarchical assembly, but the parity ends there.

C-command does not account naturally for the phenomena it is supposed to account for. Nothing indicates why it should play a role in binding. In contrast, priority and finality bind for good reasons. Anaphors are referential, but not independently so. They have to acquire reference through association with some fully referential noun. When they enter a structure, they seek such a noun. Economy considerations dictate that the nearer that noun is, the better; in other words, they will acquire reference from the nearest noun that matches features (gender, number, etc.) with them. Ideally that noun will already be in the structure—a prior attachment. If not, the anaphor must still find its associated referent before the clause is closed; otherwise reference will be impossible to establish. A final attachment thus serves as a kind of last recourse. If, as in (24), features don't match those of a final attachment, the sentence becomes ungrammatical:

24. *Mary mailed a letter to themselves.

As regards the third factor that argues in favor of the present theory—avoidance of complex and/or unnatural devices—the principle of priority removes the need for Larsonian shells, quantifier movement, and the covert operations these involve. Indeed covert movement should be eliminated entirely from any grammar that aims at neurobiological plausibility. It has been used quite recklessly in generative models, partly because of its convenience in "solving" problems like those presented by the Barss and Lasnik (1986) data, partly because its implications for any realistic model of how the brain actually produces sentences have never been debated. The latter fact may have been excused by the competence-performance distinction and consequent focus on competence. But the shift from competence/performance to the i-language/e-language distinction clearly obviates this strategy—the way sentences are put together forms part of i-language (it is internal and individual) and therefore must be integrated in any overall description of i-language. Meanwhile the fact that the problems resolved here arose not from empirical data but from a theoretical assumption—that binding must involve c-command—should not be allowed to pass unnoticed.

Island Effects

First noted by Ross (1967), island effects restrict the domains from which extraction is possible. Take the following examples:

25. What did Bill think [Mary had not properly explained___]?
26. *What did Bill wonder [who explained___]?
27. *What did Bill deny [Mary's claim that she had explained___]?
28. *What did Bill get annoyed [when Mary tried to explain___]?

While *what* can be extracted from its original position in the subordinate clause in (25), it cannot be extracted from a similar clause in (26), from a complex phrase in (27), or from an adjunct clause in (28). Examples (26)–(28), plus some other environments (e.g., subject NPs and coordinate structures), are said to constitute "(syntactic) islands," a topic that has been extensively discussed in the literature (e.g., Chomsky 1981, 1986; Boeckx 2008) but is still far from resolved.

Since Ross (1967) uncovered them, island effects have usually been explained by the principle of subjacency. Earlier versions of subjacency (e.g., Chomsky 1986) regarded it as involving "bounding nodes" that served as barriers to movement; later (minimalist) versions invoked some version of "Shortest Move." Whatever the mechanism, the aim was to block movement, mainly involving question words, out of more deeply embedded clauses, as shown in examples (26)–(28). However, Belletti and Rizzi (2000: 27) quote Chomsky as saying that "there is no really principled account of many island conditions"; a decade later "there is still no principled characterization of islandhood" (Progovac 2009: 307). Progovac attempts an ingenious solution that stands the problem on its head: traditionally movement has been regarded as the norm and blockage of movement the datum that requires explanation, but Progovac reverses this and treats movement as the extraordinary case, based on her conclusion that "syntactic islands do not form a natural class" (328).

In fact under the present analysis syntactic islands do form a natural class, and a principled account of island effect is thus possible. Stated informally, the leading idea is that while movement is normally possible to the left periphery of any clause, further movement is possible only if the home clause of the moved constituent has not been closed. The notion of closure is crucial to the present model. Though I don't know of any other version of generative grammar in which the notion is used explicitly, it is implicit in several treatments and comes nearest to what is proposed here in the notion of phases (Chomsky 1999). We may define closure as follows:

29. A syntactic structure X is closed if X is not an obligatory modifier of V.

"Obligatory modifier" includes (and may be limited to) subcategorized arguments of verbs. Note that here I do not use the formula for "if and only if" because while (29) is probably all that was included in UG, other ways of closing structures subsequently became available, as shown in Chapter 6.

It is assumed throughout that what is here (following a long generative tradition) informally described as movement actually consists, as in much recent work, of copying an already attached constituent and subsequently deleting the copied item. This is illustrated in the examples that follow, where a word enclosed in angled brackets ("less than" and "greater than" signs < and >) is not pronounced but indicates a prior position of a copied constituent. Note that as a consequence of this, a "moved" constituent is always bound by a prior antecedent, showing the ubiquity of the priority condition on binding.

30a. Mary Bill saw <Mary>, but not Sue.
   b. Who did Bill see <who>?

These examples raise no problems; only one clause is involved.

31. The man [you saw <the man>] wasn't Bill.

Here the bracketed clause modifies a noun head, but movement is not involved: *the man* is attached in both clauses but remains unpronounced in the subordinate clause. The two are understood as coreferents, but binding is not involved.

32. What [did Mary think [<what> (that) John had done <what>]]?

The interior bracketed clause is a Theme argument of *think,* therefore meets the definition in (29) and consequently remains open. This relationship does not obtain in (33):

33. *Who did Mary go home [after she had seen <who>]?

Here the bracketed clause is a modifier of the higher verb *go* but is not subcategorized by it and is therefore not obligatory; consequently the adjunct clause is closed.

34. *Where did [the fact [that Mary went <where>]] surprise John?

Here the clause inside the inner brackets modifies not a verb but the noun in the outer brackets, leading to closure.

35. *What does Mary play tennis and [John plays <what>]?

Here the bracketed clause does not modify anything; both clauses are complete and hence closed.

There are other cases that seem at first sight to be counterexamples to the obligatory modifier condition:

36a. *Where is [that Mary went <where>] likely?
  b. Where is it likely [that Mary went <where>]?

This is a classic case of what have been termed "subject islands." But subjecthood per se cannot be what is at issue here, because in other contexts extraction from subjects is licit:

37. [Who [Bill knew <who>]] shocked all of us.
38a. Mary says [who Bill spoke to <who>] doesn't matter.
  b. Who Mary says [<who> Bill spoke to <who>] doesn't matter.

In (37)–(38) a subcategorized argument is involved, so free extraction is predicted. But this is equally true of (36). The only difference is that while *that* (absent, note, from (37)–(38)) may be omitted from (36b), it may not be omitted in (39):

39. *Mary went home is unlikely.

This suggests that *that* in (36a) is different from (or is performing an additional function to) *that* in (32) and (36b). *That* is of course a polyfunctional word, serving as a determiner *(that book)*, a demonstrative pronoun *(that is the book)*, a relative pronoun *(the man that left)* and a complementizer *(I think that that is the book)*. There is yet a fifth function, that of closure. That this function is distinct from that of complementizer is shown by examples like (42):

40. I believe that that Obama is still president is indisputable.

The first *that* in (40) is a factive-clause complementizer, but the second performs none of the first four functions listed above. In fact it is a pure marker of closure.

Though too clumsy for normal utterance, (40) is (at least by my judgment) fully grammatical and merely a contraction of the slightly longer but less clumsy (41), which is unquestionably grammatical:

41. I believe that the fact that Obama is still president is indisputable.

Compare also (42) with (43):

42a. That Mary went home is inconvenient.
  b. It is inconvenient that Mary went home.
43a. The fact that Mary went home is inconvenient.
  b. *It is inconvenient the fact that Mary went home.

One can say (44):

44. It is inconvenient, the fact that Mary went home.

But (43b) is simply a dislocation of (43a), not an equivalent of it in the sense that (43b) is an equivalent of (43a). (44) is therefore not a counterexample.

That closure rather that subjecthood per se is the critical factor here is confirmed by a further example in which the subject clause does not contain *that* (Kluender 2005: examples 9 and 10):

45a. *Who does [that she can bake ginger cookies for __] give her great pleasure?
  b. Who does [being able to bake ginger cookies for __] give her great pleasure?

(45b) is slightly awkward but far more acceptable than (45a). Consequently it would appear that the nonextractability of items in subject islands is not due to a property of such islands per se but rather relates to whether or not those islands have been closed. More will be said about closure in Chapter 6. For the moment, it suffices to note that the initial generalization—that movement may take place out of main clauses and subcategorized subordinate closes—appears to be a valid one.

The Reference of "Empty Categories"

The phenomena described as "empty categories," "null constituents," or phonetically unexpressed phrases (Jelinek 1984; Huang 1984; Lasnik and Uriagereka 1988) present a serious problem for empiricist accounts of language acquisition. How do children learn inductively the reference of constituents that are inaccessible to any sense? To the best of my knowledge, no empiricist has dealt seriously with this issue. Rare attempts at doing so involve nothing more than notational tricks, and even these only address cases such as WH-questions and passives (see, e.g., Pickering and Barry 1991), where the relationship between gap and referent is fairly transparent. They certainly never deal with cases like the following:

46a. Mary is too angry___ to talk to___.
  b. Mary is too angry___ to talk to her.
47a. Mary needs someone___ to work for___.
  b. Mary needs someone___ to work for her.

Here nothing has "moved," so there is nothing one can trace back to its "original" site. Moreover there is the additional complication that in the (b) sentences replacement of the second empty category by an overt pronoun changes both its own reference and the reference of the first empty category.

Suppose we simply treat these as we treated binding of overt anaphors, that is, by the principles of priority and finality, with priority applying first, if possible, and finality supplying the default referent. The process is algorithmic, starting from the most deeply embedded gap, which in both (46a) and (47a) is the second. In (46a) this gap fails to find a prior antecedent so goes to the final attachment in the structure headed by *talk to* and finds that this is also a gap, therefore unable to supply reference. Thence it proceeds to the final attachment, which is *Mary,* a referential item. Since there are no further referential items, and a gap, unlike a pronoun, cannot look outside the sentence for reference, the first gap must remain without reference and be interpreted as generic—"anyone"—as is shown by the fact (46a) is synonymous with its gapless equivalent (48):

48. Mary is too angry for anyone to talk to her (Mary).

In (47a) the second gap does find a prior antecedent: *someone.* Although *someone* is not a definite referential, it is still referential, so the second gap refers to it. The first gap can now refer to the final attachment, *Mary.* Note that in no case can two gaps in the same clause refer to the same item. All participants in any action or event are assumed to be distinct from one another unless their identity is marked explicitly by a reflexive pronoun, a reflexive marker on the verb, or some similar device. What in the mind/brain makes this assumption a default one remains mysterious.

Replacement of the second gap by a pronoun changes the reference of remaining gaps, in part because pronouns have a different algorithm for establishing reference and in part because changing the sequence of gap-reference assignment inevitably changes reference. In (46b) the first and only gap can now take final *Mary* as its reference. But as a result, *Mary* becomes the notional subject of *talk to,* and pronouns cannot take clause mates as antecedents. In consequence, *her* must now refer to some unspecified female other than *Mary.* In (47b) the first gap is now first to search for a referent and immediately encounters prior *someone,* which gives it reference. Since *someone* is now the notional subject of the lower V, *her* is no longer excluded from taking *Mary* as its referent.

The principles illustrated here are not specific to English. They can be found in a variety of languages and in constructions of quite different types. For instance, they apply to serial verb constructions (Sebba 1987; Baker 1989). Serial constructions are highly plausible candidates for having been more frequent in much earlier stages of language, and their existence follows logically from a reasonable surmise about those stages. When only the basic universal algorithms for building phrases and clauses existed, we may suppose that there were no restrictions on what could qualify as a modifier. In that case, there was nothing to prevent a verb from modifying another verb. However, in so doing the second verb would not be a head and would therefore be unable to take modifiers of its own (and modifiers, in this treatment, include arguments of a verb). The gaps left by subcategorized arguments are then assigned reference in exactly the same way as in examples (46) and (47).

Consider the following examples, the first two from creole languages, the third from an African language:

49. mi suti   en kii (Saramaccan).
    I shoot him kill.
    "I shot him dead."
50. li pran ti   lisyen tue (Seselwa).
    he take small dog   kill.
    "He killed the little dog."

51. A'sı'ba' be´ le'sı' gu' (Gungbe; Aboh 2009).
    Asiba collect rice eat.
    "Asiba ate a lot of rice."

In all three cases, the empty objects of the verbs *kii, tue,* and *gu* take prior attachments (the main-clause direct objects *en, ti lisyen,* and *lesi,* respectively) as their referents, while the empty subjects of those verbs take reference from root-sentence final attachments *mi, li,* and *Asiba,* as predicted by the principles of priority and finality.

Does the algorithm for determining the reference of empty categories form part of an innate, language-dedicated mechanism? Or is it merely a domain-general problem-solving device that just happens to exploit relationships that *are* part of that mechanism? Since the mechanisms (priority and finality) that underlie the assignment of reference here are identical with those that assigned reference to anaphors and "moved" items, since those mechanisms specify relationships between constituents that must exist in any grammar that uses a Merge- or Attach-like process to build sentences, and since the search space is delimited by another element of the present model, closure, we can regard the algorithm that determines reference as a part of UG that follows logically from Attach, Close, and the phrase and clause algorithms.

After Universal Grammar

Attach, Close, and the phrase and clause algorithms constitute, on the present account, the totality of Universal Grammar in the sense of specific computational mechanisms for generating syntax. There are also things that are vital to language that result, apparently, from certain dispositions of the brain whose origin is obscure. We will encounter some of these in Chapters 6 and 8, but they involve words and their meanings rather than structure. The ingredients of UG itself seem both minimal and natural. Without some such process as Attach, communication would never escape the protolanguage state; not only would there be no complex sentences, but there would also be no complex thoughts, since these require syntactic structure just as much as sentences if they are to escape exponentially growing ambiguities. Closure is an economy measure from more than one perspective. Time, in both thought and communication, is saved if clauses are constructed in parallel rather than serially. Effort is saved if the space over which the brain must compute at any given time is reduced, with completed sections of a complex sentence being put to one side, so to speak, so that they no longer enter into the computation.

If words fall into two major categories, those that concern entities (nouns) and those that concern actions, events, or states (verbs), and if single words have to have others added to them, then phrases headed by nouns and clauses headed by verbs become inescapable, and simple algorithms for producing them must be developed. Those algorithms involve no more than serial Attachment, and from serial Attachment, in turn, two relations directly fall out: Priority (X is attached before Y) and Finality (X is the last referential item to be attached to a clause or phrase). These relations in turn serve to specify what items can refer to one another, as in the binding and empty-category cases. Priority in particular, as shown by Pulvermuller (2002), is something the brain can automatically keep track of; mechanisms originally adopted for movement detection can reliably indicate the order in which words are assembled.

In addition there has to be, in any version of UG, a syntactic displacement rule that permits any constituent of a unit that has not undergone closure to be reattached at the left periphery of that unit, an operation that can apply successively wherever appropriate conditions are present. Given these ingredients and nothing else but an adequate lexicon, it is possible to generate sentences of the level of complexity shown in (52)–(54):

52. Large black dog bite people dog think be dangerous.
53. Mary want know who kill antelope big forest yesterday.
54. What John say John think new tribe go do?

I am assuming that pronouns were not an early invention and that, apart from a small number of adjectives, the lexicon still contained nothing beyond nouns and verbs and a handful of modifiers, possibly undistinguished by class. (These assumptions are not in any way essential for the theory.) Even with these limitations, a child in that period would not have had to do more than learn vocabulary in order to acquire language. The rest would have been provided by automatic processes already established in the nervous system. You will note, however, that the examples leave something to be desired in at least two respects: they are less than clear on where one phrase or clause ends and the next begins, and who did what to whom is not always as obvious as it might be.

This is because the UG presented here is radically underspecified. There are whole areas of grammar, such as word order, agreement, and functional categories, on which it remains silent. This might not be what one would expect from the last half-century of theorizing about UG, but it is exactly what one would expect from an evolutionary adaptation. Evolution cannot predict the requirements of something that, once established, might turn out to have those requirements. It can only read what it sees. What it could see was that a species was using symbolic items by stringing them together, and it set about accelerating, expanding, and automatizing this process, as it had automatized many others that had proven adaptive (bipedal running, aimed throwing, Acheulean hand-ax knapping, etc.). Although no specific aspect or device of language was in and of itself adaptive, the whole package, at any stage of development, was calculated to enhance fitness. Consequently genetic factors would ensure that each development would spread through the population. What they could not do (or did not have time to do, if the out-of-Africa diaspora intervened at this stage) was fine-tune the linguistic engine, particularly in order to meet the needs of the hearer.

UG itself is usually regarded as being neutral with respect to speaker and hearer. It should make it easier (automatic, in fact) for speakers to produce sentences, while also making it easier for hearers to process sentences than it would have been if they had remained unstructured. The question is, did it make it easy enough for the hearer? Some of the things UG failed to spell out—relative positions of head and modifier, for example—could be resolved by conventions of use. Others—determining who did what to whom, what constituents were closely linked structurally with one another and which were less closely linked, where one phrase or clause ended and another began—were not completely spelled out or not spelled out at all.

The existence of underspecifications in UG is precisely what leads to the wide degree of variation and the constant processes of change that are found throughout the world's languages. This chapter has dealt with what cannot vary or change. At least a large proportion, perhaps even all nontrivial types of variation and change, arise from a single factor: the need to specify things UG left unspecified, a need driven mainly by problems the brains of hearers had in processing the bare output of UG.

Before we proceed, it is important to note that any shortcomings of this restricted UG with regard to communication did not apply when it was used for thought. Indeed, that is not going far enough. What was a communicative disadvantage was actually a strong cognitive advantage. For offline, "abstract" thinking, all that was required was that concepts be linked with one another. The clutter of case and agreement and all the other variable features of language could only be an impediment to thought. Thinkers normally know what they are thinking; they don't need signposts to sort it all out.

Bear in mind always that the brain is neutral as regards cognitive and communicative functions. Both involve the same algorithms. Both link together clusters of neurons that have representational functions (whether words or concepts) by the same means. If communication is involved, further processing is necessary; that is the only difference. That difference explains why, though thinking is the same the world over, language varies significantly from place to place.

CHAPTER 6

Variation and Change

We have now effectively finished the part of this book that deals with biological evolution. The human brain, once inseminated with words, provided an instrument that enabled members of the species to acquire and use language. After that it was up to those members to make what they could of it. Of course, they could use linguistic materials only in ways those materials permitted and could change them only by adding features, not by altering or discarding fundamentals. It follows that subsequent developments, while constrained by UG, were primarily cultural innovations, and it follows too that they would have to be learned inductively. Bear in mind that one of the things we have to explain is why, if human cognition is uniform worldwide, human languages appear to be so diverse.

This approach, one based on a principled distinction between unlearned (biologically based) and learned aspects of syntax, differs radically from both empiricist and nativist approaches. For empiricists, all of syntax is learned inductively. For nativists, only trivial, peripheral aspects of it are learned inductively. Consequently the bar for justifying such a theory is higher than it would be for any theory that fell squarely within the terrain of either the nativist or the empiricist camp.

Such an approach has to show not merely that it fits the facts of language but that it can illuminate those facts in ways that neither alternative can. It should be able to make sense of things that other approaches failed to explain or simply ignored. It should lead to insights that in turn yield new research goals. It should be able to do this most explicitly in three areas that have in various ways proved difficult for existing theories of language to deal with. Those three areas are language variation and change, language acquisition, and creolization and its resultant languages, which unites aspects of both. Accordingly the present chapter deals with variation and change and Chapters 7 and 8 with acquisition and creolization, respectively. In Chapter 9 we return to Wallace's problem.

Why Variation and Change?

Chomsky (2010: 58) saw two questions about language as salient. The first was why there are any languages at all—after all, the world had remained without language for four billion years, and countless millions of species had lived and died without any of them developing one. Chapters 3 through 5 dealt with that issue, showing that only a two-stage process—an ecologically driven breakout from the restrictions of prelinguistic communication, followed by a brain-driven enhancement of behaviors that this breakout made possible—could have given an otherwise not particularly distinguished primate powers so much in excess of its needs. The second question was why there are so many languages (and, one might add, why those languages seem to differ from one another as much as they do). We can speak about the song of Bengal finches or the communication of vervet monkeys as if entire species shared the same code. Why can't we similarly speak of the language of humans?

In the words of Baker (2001: 207), "Why doesn't our innate endowment go the whole way and fix all the details of the grammar of human language?" Baker agrees that one frequent answer—because language adapts itself to the requirements of different cultures—won't work. There are no consistent correlations anywhere between any cultural feature and any linguistic feature. Typologically similar languages can be found among sub-Arctic herdsmen and dwellers in tropical forests, but you can also find a range of very different languages in each of these culturally relatively homogeneous groups.

In fact the arguments of Deacon (1997) and Christiansen and Chater (2008) apply here. Once humans had the materials for a starter language, change was inevitably going to take place at a rate too fast to form a target for natural selection. Take a case where human behavior surely *has* been incorporated into the genome: the precision, flexibility, and strength of the human handgrip. We inherited a good basic grasp from branch-gripping primate ancestors, but try to imagine a chimpanzee texting on an iPhone and then repairing your watch (even if it knew exactly what it was supposed to be doing). Humans have developed the primate grasp into two basic grips, a power grip and a precision grip, arguably from the actions of throwing and clubbing, respectively (Young 2003). But these are actions that were performed in similar ways over a long time period rather than in different ways over relatively short periods, as must have been the case during the development of language, once its genetic component was in place.

Moreover once basics were in place, further specific developments were no longer evolvable. No grammatical structure in and of itself confers any adaptive advantage over others. Skill over language as a whole—faster speech that made more sense, processing that extracted that sense more rapidly and accurately—was the only thing that might confer enhanced fitness. Moreover, as we saw in Chapter 5, even UG itself was at least as much due to the brain's own need for economy of wiring and automatic, repeatable routines as to natural selection sensu stricto. Once the brain had those two things, it had all it needed. Why would it have gone further? And how, even if it had, would it have generated not a set of species-wide features but a flood of variables?

However, answering Baker's (2001) question in this way merely poses another question: If by the time the human species emerged (or shortly thereafter) we had a genetic recipe for language, why did language need to develop any further? As far as a language of thought is concerned, that recipe probably hasn't changed or been added to since the basic algorithms for syntax first appeared—and we know how effective it is for that purpose. As a means of communication, it was orders of magnitude more effective than the communication of any animal, or than its preceding protolanguage, for that matter.

This follow-up question has no single answer. A variety of factors contributed to the current state of affairs, factors that first operated alone and later in concert to produce variation and change. Bear in mind that these two things are fundamentally the same. Change is simply variation on a temporal axis, as first made fully explicit in Weinreich et al. (1968) and later developed in work by Labov (1972) and colleagues. Consequently we can treat both as different aspects of a single phenomenon:

what must happen when you have a UG that supplies the major building blocks for language, but by no means all of the building blocks that language users might find useful or even indispensable.

To the best of my knowledge, no one has looked at language change and variation in quite this way. Previous studies have concerned themselves mainly with classifying and describing different kinds of change. Some recent studies have focused on the distinction between "internally motivated" and "externally motivated" classes of change (Jones and Singh 2005; Hickey 2010). Early generative studies such as King (1969) naturally saw change through the prism of grammatical rule systems and divided changes into cases of rule addition, rule loss, rule reordering, rule simplification, and restructuring of underlying forms. Equally naturally, later generative grammars saw change as "essentially a random walk through the space of possible combinations of parameter settings" (Battye and Roberts 1995: 11). Ayoun (2003: 12), who notes the survival of parameters into the Minimalist Program, states this position even more flatly: "Adopting a different value for a given parameter could be the only way a language may change over time."

As the foregoing suggests, students of language change have been much more concerned with asking *how* languages change than in asking *why* languages change. Over two decades ago, Breivik and Jahr (1989: 1) "felt it safe" to stake the extremely modest claim "that historical linguistics has now left the stage where *all* the causes of language change are unknown" (emphasis added). But the essays in that collection all deal with causes of particular changes in particular languages. and no general theory of linguistic change is even suggested. Fifteen years later one source concluded that "although the prestige of a particular variety plays an important part, language changes can never be predicted" (Dirven and Verspoor 2004: 204).

Consequently while the overall landscape through which this chapter will travel may seem familiar enough in its details, the treatment will not. To break a new trail through even a well-trodden landscape (and, as suggested earlier, the landscape of diachronic linguistics may be well-trodden but is hardly well-mapped) should inevitably present novel views of the scenery. Hopefully some of these may help to show why, when it comes to change and variation, languages behave the way they do.

Inherent Instability

The first factor to look at is perhaps the most ineradicable: the inherent instability of the phonetic element. Phonetic space—the area within which speech sounds can be produced—is, because of the structure of the human vocal organs, without clearly marked internal boundaries. This is most obvious in the case of vowels, but consonants are also affected. Consonants may have, in principle, distinct places of articulation, but in practice these vary due to a variety of influences, such as whether a sound is pronounced at the beginning, middle, or end of a word, or what other sounds occur in its immediate vicinity on any given occasion. All speech sounds are, objectively speaking, gradient, but human perception of speech sounds is categorical. What this means is that although it is impossible on acoustic evidence to draw nonarbitrary lines between any two adjacent sounds, humans will inevitably hear either the "same" sound twice or two distinct sounds. That is at least in part because it is contrasts between sounds just as much as the sounds themselves that enable us to recognize a word and distinguish it from similar words.

Phonetic instability creates a pool of variables that can then be pulled in different directions by a variety of extralinguistic factors: social, cultural, or merely statistical. Chance fluctuations in the frequency of particular variants can reach a tipping point and precipitate categorical change. Some pronunciations become fashionable, perhaps because the group using them acquires prestige, while others, used by disadvantaged groups, are demoted to an underworld of "nonstandard" pronunciations. Nothing better illustrates the sheer arbitrariness of this process than the social fate of the two phonetic variants of the past tense of *eat* (/eyt/ and /et/) in British and American English, respectively. In the United Kingdom /et/ is the upper-class and /eyt/ the lower-class pronunciation; in the United States the status of the two is reversed.

The effects of phonetic instability are not limited to phonology. Ripple effects spread instability to other linguistic levels. Changes in pronunciation can make words unrecognizable in a few thousand years, so that even if a language was determined down to its last details by genetic factors, its lexicon would still be subject to change. Rapid and casual speech, by far the commonest kind in small face-to-face societies, accelerates and intensifies changes arising from other causes. This affects all words up to a point, but words with inflectional prefixes or suffixes suffer the most. Grammatical information tends to be expressed at the beginnings and especially at the ends of words, but it is precisely here that phonological instability has most impact. Sounds within words change relatively slowly, because they are always pronounced (more or less) the same way. But sounds on the peripheries of words contact a wide range of different sounds from the beginnings and ends of immediately adjacent words. Consequently peripheral sounds are often changed or lost altogether, causing ripple effects that spread through grammatical morphology to syntax, as when (for example) suffixes expressing tense on Latin verbs became indistinguishable or disappeared, causing Latin's daughter languages to replace them with periphrastic expressions employing verbs like the equivalents of "have" or "be."

But there were other factors that contributed to variation and change, perhaps the most important being that UG was radically underspecified in two quite different ways. One kind of underspecification existed because there were things unspecified in UG that had to be specified in speech. The other kind were cases where additional specification, though not strictly necessary for communicative purposes, was seen as enhancing the efficiency of communication.

Underspecifications That Must Be Specified

There are at least two major areas for which no structure is specified in the version of UG described in Chapter 5: word order and action, event, and state descriptions.

Word Order

The reason word order seems so important in language might at first sight be taken as stemming from the fact that language uses only one output channel at a time. It can use speech or manual sign, and there are people who can simultaneously speak and sign the same message, so it is possible to imagine a human language that incorporated elements of both—one, for instance, in which all nouns were spoken but all verbs signed. Such a language would save a lot of time, because nouns and verbs could be expressed simultaneously. But no such language exists, or (so far as we know) ever existed. So the reason for the importance of word order is probably less a shortage of output channels than a condition on processing: we can process only information that is presented sequentially.

But whatever its motivation, sequential utterance forces us to specify word order. A rough word order falls out from simple attachment of closest constituents—verbs always start an attachment process, objects are (usually) attached before subjects, and so on—but relative placement of heads and modifiers is nowhere specified in UG. Because language is linear and sequential, the choice must be made. But because the brain, far from designing an optimal language, is merely satisfying its own needs for wiring economy and automated routines, no instructions have been provided on how to make that choice.

There is a tendency for either head-first or modifier-first strategies to involve many if not all structural categories. If in some language verbs precede their complements, then in many and probably in most cases nouns will precede their adjectives. But there are numerous exceptions (English is one) even to this generalization. Grant that exceptions may often result from linguistic change, as this one does; the very existence of such change argues against any principled, across-the-board language universal involving head-modifier ordering.

In the early Middle Ages Old English changed from an O(bject)-V(erb) order to a V(erb)-O(bject) order. Expressing a standard generativist view, Lightfoot (1991) has described such changes as resulting from a series of exceptions to OV order accumulating until the basic structure of the language was quite abruptly reanalyzed as VO. This view has been specifically endorsed for Old English by Koopman (1990) and Stockwell and Minkova (1991), among others, but Pintzuk (1996, 1999) showed from contemporary sources that there was no sudden reanalysis but rather a long-drawn competition between alternative grammars with OV and VO orders respectively. That competition didn't end decisively. Adjective-noun order in English still follows the head-last pattern of OV languages, unlike the Romance languages, which, although descended from (predominantly) OV Latin, have completed the head-modifier switch for all categories.

How, then, to account for the frequency of uniform head-modifier ordering? It would seem to be due to an extralinguistic factor: a preference for symmetry, widespread among humans, that has been shown in fMRI studies to be supported by intensified brain activity (Sasaki et al. 2005). As for the inherent instability of ordering, particularly as it involves the categories of Subject, Verb, and Object, one need look no further than the fact that there is no known language that adheres always to a single categorical ordering. All languages allow themselves some freedom in the distribution of these major categories, since speakers of all languages have communicational needs that involve the foregrounding or backgrounding of information. All it takes is for sentences with noncanonical ordering to occur so frequently that some speakers begin to treat them, first, as alternatives to, and later, as replacements for the language's original canonical ordering.

Tense-Modality-Aspect (TMA) Systems

The second area of underspecification that must be repaired covers the expression of the relative timing, reality, or degree of completion of particular actions, events, or states (in terms of grammar: tense, modality, and aspect). Note that in a few languages (e.g., Chinese) not all these categories are obligatorily marked, but two things should be noticed. First, there is a good deal of overlap in the categories: if an action is +completed (aspect), it must be +past (tense) (except for the so-called future perfect, e.g., *will have left*). Second, even where marking is not obligatory, there will always be optional means of expressing the category. So possession of some kind of tense-modality-aspect (TMA) system is a presumptive candidate for universal status.

However, the absence of any universal pattern for expressing those categories, indeed the very wide variety among both categories chosen and means adopted, argues against that status.

Means include (but are not necessarily limited to) inflections on the verb and free particles, which may be obligatory or optional and in either case may precede or follow the verb. In the simplest conceivable model, there would be just one marker for tense (past versus nonpast), one marker for modality (things that had not yet happened or might never happen, i.e., were unreal, vs. things that had happened or were

Variation and Change   159

still happening, i.e., were real), and one marker for aspect (ongoing or repeated actions vs. single, punctual ones). The zero form (the verb stem, without marking) would do duty for the unmarked (most expected, most frequent) case: past for tense, real for modality, completed for aspect. Note, however, that "past" may be computed from two reference points: the present moment (the norm among European languages) or the time of the main topic of current discourse (the norm among creole languages and many West African and Southeast Asian languages, usually referred to as "relative past").

The TMA system found in many creole languages comes closest to this model (Bickerton 1981, 1984b), and for reasons discussed in Chapter 8 I originally assumed it must be specified in UG. I no longer believe this to be the case. There seems to be a very clear distinction between things specified in UG (the processes Merge or Attach and Close, the phrase and clause algorithms, and the relations that fall out from these) that are present in all languages, and things left unspecified (word order, agreement, case, and much more) that vary between languages. The TMA system is one of these variables; indeed (outside creoles) it is hard to find any pair of languages that have identical systems. Thus English speakers often have a hard time getting their heads around the category "subjunctive," even though subjunctive is no more than a subdivision of the modality category [-real].

Yet the ubiquity with which this simplest conceivable model, or something closely approximating to it, is found among creoles argues that, even if TMA is not specified in UG, there is a strong bias, resulting perhaps from some mix of perceptual, memory-based, and other nonlinguistic factors, in favor of adopting it whenever evidence for any competing model is absent or ambiguous. In other words, the categories of tense, modality, and aspect themselves (as distinct from their varying instantiations in different languages) appear to arise from some way of dividing up reality that likely precedes the appearance of language. Note that the basic creole system (stem plus one marker each for tense, mood, and aspect, yielding eight different structural forms) represents a combination of the minimal structural apparatus with the maximum semantic coverage.

TMA systems worldwide vary on a number of axes. One of these is the semantic scope of the system. Some languages have TMA systems that incorporate semantic notions additional to the time/reality/event-type-based trio that dominates TMA systems generally. In a number of languages there is something called "evidentiality." Evidentiality encodes such things as how the information in a sentence was obtained—whether from the speaker's own experience (in which case the sensory channel may be noted), from hearsay (in which case the status of the informer as witness or nonwitness of the events observed may be explicitly marked), from plausible inference, from speculation, or from generally accepted knowledge—and whether it was obtained recently or in the more distant past. Most languages that have evidentiality encode only a couple of these distinctions, though in extreme cases (e.g., Matses, an Amazonian language; Fleck 2007) a language may include most of them. This is an example (we will meet more) of a feature either necessary or useful to language but not specified in UG that may therefore take on a life of its own.

Another way TMA systems can vary is increased specification. For example, languages may divide the category [+past] into immediate, recent, and remote past. They may divide the category [-real] into wholly imagined and possible future events, with further divisions based on speaker intentions for the future and/or the relative likelihood that the events described will in fact take place. They may divide the category [-completed] into actions that are continuous (still in progress at the temporal reference point) as opposed to actions that are discontinuous but still being repeated (or likely to be repeated) on both sides of that reference point (e.g., *Mary walks to work every day*, though she's not doing that right now). Overall (if we exclude nonuniversal components such as evidentiality) any randomly selected pair of TMA systems will probably divide an identical semantic area into a different number of categories. It follows that none of those categories will have exactly the same meaning or function in both languages, just as, if a pie is cut into seven slices, at least some of those slices will differ in size and content from those of the same pie cut into ten slices.

If TMA systems show this degree of variability, how are we to account for the universality of the three basic distinctions of past/nonpast, real/unreal, and completed/uncompleted? To establish the first distinction, even if one has adopted relative past, this needs only an episodic memory to store events in a serial order that is permanently recoverable.

Humans do indeed have an autobiographical memory (Conway 2005; Conway and Pleydell-Pearce 2000): we can almost always recall the sequence in which things happened to us, even when we are unsure of precise dates, so a relative past should present no problem. To establish the second, all one needs is for the brain to be able to differentiate between stored memories and things for which no memory is (yet) stored. The basis for the third distinction is less clear (as is its importance), but that may be merely because we still know so little about how memory (or much else in the mind) really works. Overall the mix of universality and variability that TMA systems exhibit suggests neither a linguistic universal nor a series of ad hoc inventions but rather some nonlinguistic universals that have been incorporated into language in varying ways.

Underspecifications That Need Not Be Specified

Inherent instability due to continuous phonetic variation (inevitable, given the kind of speech organs we have) plus the underspecifications of UG in terms of its lack of detailed instructions about relative positions of heads and modifiers and its failure to specify structures for nonlinguistic universals made it inevitable that full spoken languages would have to vary and change. But things could have stopped there. A language that had selected its word order (with or without obedience to universal symmetry) and found some means to express TMA categories would have as much expressive power as any modern language. Indeed the existence of languages such as Riau Indonesian (Gil 1994) suggests that nothing in principle prevented this. What is not clear is "why all of us should have the capacity to learn natively languages whose units include not only phrases and sentences" (Carstairs-McCarthy 2011: 436) but all the other complexities we find in grammar. Still less clear, in light of Riau Indonesian, has been why those complexities exist at all. We have had many studies of *how* languages change, in particular studies of grammaticization, the process that takes content words and makes them into grammatical markers that may later become attached to larger words (Heine and Kuteva 2007, 2012; Givon and Shibatani 2009). We have had very few studies of *why* such complexities might have been introduced.

The picture changes once we approach this issue from the perspective of an only partially specified UG. Protolanguage evolved to make possible recruitment for confrontational scavenging. But language (or rather that part of language instantiated in UG) did not evolve in response to any particular human need. Rather it evolved to improve the brain's speed and accuracy in processing words and concepts. It was successful not to the extent that it improved human fitness but rather to the extent that it satisfied the brain's need for economy and automaticity. Neither brains nor individuals could have foreseen that the first words that grew into protolanguage would eventually "provide a uniform format for all concepts" that "mixes conceptual apples and oranges" (Boeckx 2012: 498). Nor could they have foreseen that the syntax that grew out of protolanguage would create the most powerful cognitive mechanism that had ever existed.

All that (at some period perhaps around 200,000 years ago) still lay in the future. What lay in the immediate present was the fact that what the brain had devised was less than optimal for human communication.

I noted in Chapter 5 that the instructions of UG were biased in favor of the speaker. They made sentences easy to create (indeed made them quite automatic and without any need for conscious intervention). Put together with this the fact that humans do not come neatly divided into speakers and hearers. This is fortunate, since speakers and hearers have an inherent conflict of interest. Speakers want to reduce to a minimum the structural apparatus they use because there is an energy cost attached. Hearers want to have speakers' messages spelled out as precisely as possible to ensure correct understanding. However, every speaker is just as often a hearer, and every hearer is just as often a speaker, so there was no real conflict of interest for those who (as speakers) found sentence easy to produce but (as hearers) considerably more difficult to parse and understand. In other words, it was in everyone's interest to spell out the relations between things represented in sentences more clearly and unambiguously than the skeletal syntax alone could do.

Language is intrinsically and inescapably hierarchical in structure, and all its significant structural relations are hierarchical and vertical rather than linear and horizontal. This is not problematic for the brain because the brain is built to create hierarchies and transform them into linear structures. The brain was not, so far as is known, similarly specialized to transform resultant linear structures back into their original hierarchical structures. It can be done, but there is a computational cost, as well as the risk of misinterpreting content. The problems in this kind of transformation are at least threefold.

First, there is the problem of determining which words are most closely related with which other words. In a hierarchical tree structure this is clearly apparent: you can see instantly what is directly attached to what, and what units form groups under a single dominant node. In a linear structure you can see none of this. For instance, take a sentence such as *Here customs that to the average educated person seem like ancient history persist into the present century.* Nothing in this sentence except its structure tells us that the whole phrase *customs that to the average educated person seem like ancient history* is the subject of *persist* (with which it agrees in number) and takes plural number from its head *customs*. That there can be confusion in cases like this, where two singular nouns intervene between plural *customs* and the verb, is shown by the frequency with which similar sentences in the present tense, even in writing, are encountered with the singular agreement marker on the verb *(customs that . . . persists).*

A second problem arises because UG has no means of showing the function of different phrases with respect to their verb—the thematic roles of their arguments, in other words—other than by word order. But word order can be effective only for two arguments, and that only if the language has SVO structure and V is transitive. In such a language we know that the subject (usually an Agent if there is one) will precede the verb, and the direct object (most frequently a Theme, what the action of the verb most directly effects) will follow it. However, a sequence of two unmarked arguments after a verb (the so-called double object construction) provides no indication of which is which. (English speakers know this only because of the alternative—a prepositionally marked indirect object, as in *I sent the letter to Mary*—which shows that in the double-object alternative *I sent Mary the letter*, *Mary* must be an indirect object.)

Worse still are cases where the verb also has a string of non-subcategorized arguments, expressing things like Means *(with X)*, Beneficiary *(for X)*, Time, Place, and so on. They can't be assigned a numbered sequence because they won't all be there in every sentence. Suppose I said *I sent box Barbara mail her sister Tuesday*. Pragmatics and semantics might be able to figure out that what I intended was *I sent a box **to** Barbara **by** mail **for** her sister **on** Tuesday*. But how could communication flourish if we had to work out the pragmatics and semantics of most sentences before we could understand them?

The third problem arises in determining where one unit (phrase or clause) of a sentence ends and where the next unit begins. For instance, in a sentence such as *The book Bill sent Mary reminded me of it*, the noun *Mary* appears between and is equidistant from the two verbs *sent* and *reminded*. Nothing in the form of the sentence tells us that there is a boundary between clauses, that *Mary* falls to the left of it, that the sentence is not about Mary doing anything, but that *Mary* is the indirect object of *sent* and bears no relation whatsoever to *reminded*. In speech, of course, intonation could clarify the issue. A slight fall in the intonation pattern on *Mary* and a slight rise on *reminded* may mark the boundary. Alternatively a slight pause between *sent* and *Mary* could indicate the (identical in words but different in structure) pair of sentences, *The book Bill sent. Mary reminded me of it* (but Bill still hadn't sent the magazines). Intonation, however, is not syntax.

A slightly different kind of confusion arises where the same word serves as both noun and verb. In a Denver newspaper I once saw the headline *Spy Charges Dog Inspectors*. Nothing in this headline indicates whether it should be interpreted with *charges* as a verb and *dog* as a noun ("Some spy has imposed a fee on (or accused) people who inspect dogs") or with *charges* as a noun and *dog* as a verb ("Accusations that they are spies continue to be leveled at certain inspectors"). Disambiguation is left to pragmatic knowledge rather than being automatically provided by, say, a particle marking the boundaries of a phrase, a word-class marker for nouns or verbs, or some marker indicating that *dog* is (or is not) a modifier of *inspectors*.

Underspecification in these areas doesn't demand specification; otherwise languages like Riau Indonesian couldn't survive. But the fact remains that the vast majority of languages have acquired devices (ones that cannot form part of UG because they vary, to a large extent unpredictably, from language to language) in the ways they spell out the following three relationships:

Variation and Change   165

A. Relations between words that are closely connected with one another.
B. Functional relationships between phrases and the verbs they modify.
C. The boundaries of phrases and clauses.

It should be noted that any given device is not necessarily limited to a single function. Some forms may discharge more than one. Additionally some forms may also contain useful semantic content or may perform some other function that helps clarify or disambiguate the meanings of sentences.

Before we examine these classes in detail, however, we should note that the means of repairing underspecified structures turn out, despite superficial variety, to be surprisingly similar in nature. Repairs in all three areas almost always, perhaps without exception, use some particle of phonetic material, which may be added to sentences in exactly four ways. These ways are created by the intersection of just two variables: before versus after and free versus bound. The before-after distinction, as we have seen, is one that is forced on language by its linear nature and by the failure of UG to fully specify sequential ordering of constituents. The free-bound distinction arises from the fact that particles can be attached directly to words as parts of those words (bound) or can simply accompany them as discrete entities (free). Due to the consequences of rapid speech, these stages are frequently sequential in the history of a language: free particles may become bound, though the reverse is seldom attested. Choices among patterns do not have to be uniform across all categories of a language or all three types of relationship. For example, particles that signal functions of arguments in English (*by, to, for,* etc.) are free and before, whereas the one particle that signals person and number on verbs (third-person-singular present-tense -*s*) is bound and after.

Where do these particles come from? Most come from the same source: what were originally content words that have been downgraded, de-stressed, often curtailed (e.g., reduced from two syllables to one, normally the stressed one), and bleached of much or all of their original semantic content. The process, known as "grammaticization," has been well studied (among many others, Heine and Kuteva 2007, 2012; Hopper and Traugott 2003) and is quite uncontroversial except for one issue:

unidirectionality, or whether grammaticization must always turn content words into function words, never vice versa. However, nothing in the present theory requires either unidirectionality or its absence. The process can often be seen in midprogress, especially in creoles and some West African languages, where, among other things, verbs meaning "give" can be seen changing into markers of dative arguments and verbs meaning "say" into markers of factive subordinate clauses (Lord 1973; Sebba 1987).

Grammaticization of Relations between Words

The relationships most often grammaticized in languages include relations between words and the nouns they modify and between a verb and its associated arguments, most commonly the subject. The first type is most familiar to English speakers as "gender," since this becomes a problem the moment an English speaker arrives in Europe. But "gender" is no different in its effects from "noun classifiers" as found in Swahili or Chinese. The only major difference is that things we call "gender" are somewhat loosely based on distinctions between "male," "female," and "inanimate," whereas noun classifiers are based on many different properties of their referents (things like "human," "artifact," and "long thin object").

In English, apart from some archaic forms ("authoress" as opposed to "author"), gender does not exist outside pronouns. However, it still has implications for nouns; for example *ship,* though not marked for gender, is implicitly feminine *(May God bless all who sail in her/*him/*it),* thereby underlining the fact that in a language with either gender or noun-classifying systems, all nouns, regardless of their meaning, have to be assigned to some class or other.

In languages that, unlike English, have a full gender or noun-classifier system, it is customary for all adjectives and determiners (articles, demonstratives, etc.) to carry some marker of class membership:

1a. Las poemas suyas que me mostraron anoche me parecian magnificas.
  b. Los cuentos suyos que me mostraron anoche me parecian magnificos.
   "I thought that the poems/stories of yours that they showed me last night were marvelous."

Variation and Change 167

In these Spanish examples, articles, adjectives, and possessive adjectives all carry the gender marking of their head, even though several words may intervene between noun and adjective. Noun-classifier systems operate in a similar way. Here, however, class agreement markers may even extend to the verb, which then take the class marker of their noun-phrase subject's head:

2. Watu wazuri wawili wale wameanguka.
"Those two good people fell."

Wa is the Swahili marker for any number of humans greater than one.

As this suggests, noun classification is closely linked with verb-subject agreement. Agreement is vestigial in English—only third-person-singular present-tense -*s* survives out of a much richer array of number and person markers on Old English verbs—but flourishes in Romance languages, as anyone knows who has ever conjugated a Spanish or Italian verb. It shows unambiguously and automatically, without any need to process syntactic or semantic structure, what is arguably the most crucial semantic relationship in the clause: who did what. However, like all the other phenomena discussed earlier (and unlike the basic structures generated by UG), languages don't have to select it.

If they do, of course, other options open up. A language with rich agreement doesn't have to show a subject overtly, so long as the referent is pragmatically obvious from the situation (*I, we,* or *you*) or has been established in discourse:

3a. Juan se fué ayer. No volvera hasta el ãno que viene.
"John left yesterday. (He) won't come back until next year."

Even where there is more than one possible antecedent for the pronoun, it may be dropped:

3b. Juan encontró a Pedro hoy. Dije que era muy feliz.
"John met Peter today. He said he was very happy."

In principle this could be four ways ambiguous: John said John was happy, John said Peter was happy, Peter said John was happy, Peter said Peter was happy. But only the first is likely to be understood, even though on purely pragmatic grounds the fourth is equally likely. This is because in languages generally subject continuity is the default case; it is assumed that the subject of a clause will be the same as the subject of the preceding clause unless circumstances indicate the contrary.

Expectation of subject continuity is the cause of another phenomenon that is not normally considered in connection with things like gender but shares with it an important function. Switch-reference markers determine whether, in any two consecutive clauses, the subject of the first will be the same as (SS condition) or different from (DS condition) the subject of the second. As in the following example from the Mexican language Zuni, the marker is often attached, as a kind of early-warning signal, to the first of the two clauses:

    4a. ho' kwayi-nan   yak'o-nna.
        1SING-NOM-exit-SS vomit-FUT.
        "I will go out and throw up."
    4b. ho'kwayi-p   Nemme' yak'o-nna.
        1SING-NOM-exit-DS Nemme vomit-FUT.
        "I will go out, and Nemme will throw up."

If Spanish had switch-reference marking, (3b) could unambiguously distinguish between two pairs of the four interpretations of (3b).

Pure agreement phenomena are not limited to the subject-verb relationship. In polysynthetic languages, the verb can carry affixes that link it to other arguments. (In (5), SM indicates subject-marker and OM, object-marker.)

    5. Njuchi zi-na-wa-lum-a   alenje.
       bees   SM-past-OM-bit-FV hunters (FV = final vowel).
       "The bees stung the hunters."

(5) is from Mohawk (Baker 2002). Note that the verb alone, *zinawaluma*, could serve as a complete sentence ("They stung them"). But even when subject and object nouns are both present, as in (5), the verb stem *lum* must be accompanied by both a subject marker, *zi*, and an object marker, *wa*.

Note that cases like (5) shade into the second category of variables: markers that show the relationship between a verb and its arguments.

Variation and Change   169

Grammaticization of Verb-Argument Relations

"Who did what?" is not the only question language is required to answer. There is also "What was done to whom (or what) with what, for whom (or what), when and where and how?" Arguments of verbs play a variety of roles, but UG provides no specific tools for the job other than order of attachment, necessary for the speaker but little help to the hearer (just one more thing to laboriously process and maybe get wrong) since, as noted in Chapter 5, no invariant order of argument attachment is stipulated. Where the language happens to attach objects to the right of the verb and subjects to the left, as in English, word order makes it possible to distinguish the two. But at least half the world's languages have verbs at the end of the clause, while a substantial minority places the verb at the beginning.

Moreover even in English, SVO ordering is not always followed:

6. The letter itself I gave to Sue, but the envelope I kept.

In cases like this, context and pragmatics make it easy to determine the sense. But imagine if the second clause had been something like . . . *but the wind the fence broke*. Would that mean "The wind broke the fence" or "The fence broke (obstructed) the wind"? In English we know that (with object-fronting) OSV is a possible order, but also that SOV is impossible in English under any circumstances, so the second meaning is correct. But while in a true SOV language subject-first is usually commoner than object-first, one cannot rely on that.

One way of dealing with this is to have an object marker either on the verb (as in (5)) or on the argument itself (i.e., as part of a case system). Case is by far the most popular solution for SOV languages. Dryer (2002) found that, out of a sample of over five hundred languages, 72 percent (181/253) of SOV languages had case systems, as against only 14 percent (26/190) of SVO languages. (Verb-initial languages came halfway between at 47 percent.) These figures suggest that word-order choices, themselves variables outside the scope of UG, are the strongest determinants in the choice of a type of marking for verb-argument relations.

Case systems vary unpredictably in the extent to which they may proliferate. Leaving aside languages that have no case at all, English is one of the most impoverished. It has no case marking on any nouns or their modifiers; overt case is found only on pronouns, where there are two (Nominative and Accusative: *I/me, they/them*), unless one counts Genitive (*my/their*, etc.). Among the richest are Finnish, with fifteen (Nelson 1998) or sixteen (Holmberg and Nikanne 1993) cases, and Hungarian, with nineteen (Megyesi n.d.) or twenty-three if you include "less productive" cases (Rounds 2009), or "17 to 27" (Thomason 2005: 16). This uncertainty reflects ongoing and largely unresolved disagreements over what exactly a case marker is or is not. It has even been argued (Spencer 2008) that what Hungarian has are not case markers at all but rather postpositions expressing semantic features rather than functional roles.

Disagreements over classification have more to do with descriptive convenience than with any substantive issue. We should recall that there are only four options for incorporating materials unspecified in UG. They can be added before or after head words, and in either case they can be attached either as free units or directly to the heads themselves. By convention, case markers are limited to morphemes attached to words; free morphemes are excluded, even if these have identical functions. Any problems with defining them arise simply because all morphemes (except for a handful, like the vowel marked "FV" in the gloss for example (5), which exist solely to satisfy phonological requirements) carry some degree of meaning, however slight or vague.

This becomes clearer if we look at English prepositions. Here one of the clearest examples of a case marker is the preposition *by*, which can indicate either of two thematic roles, Agency *(arrested by the police)* or Location *(sitting by the door)*. In the case of Agency, *by* has no competitors and is probably the nearest thing to a pure case marker that English has. In the case of Location, it shares that function with a host of other prepositions—*in, on, at, under, beside, above*, and so on—and more closely resembles the problematic Hungarian markers. We should abandon attempts to force grammatical markers into traditional, watertight compartments and accept that there are functional clines with overlapping categories, where any given marker may fulfill one or several functions. Since the purpose is everywhere the same—to facilitate the receiver's analysis of incoming sentences—this seems a legitimate approach, but even to begin to execute it in detail would take us far beyond the scope of this book.

So far we have treated the functional relationships between verbs and their arguments as if these constituted a single class. In fact they represent two closely related and overlapping but still distinct classes: grammatical functions (subject, [direct] object, indirect object, etc.) and thematic functions (Agent, Patient/Theme, Goal, etc.). The mapping between these classes is far from straightforward, one-to-one mapping. If there is an Agent, it will usually be a subject, but by no means always. There will almost always be a Theme or Patient, and it will most often be an object, but if it is the only argument (often the case with intransitives) it will be the subject (in English at least) because a sentence must have a subject. But where there is a lack of fit between categories there are inevitably problems for any system of case marking. This gives rise, among other things, to the phenomenon known as "ergativity."

Ergativity provides another axis along which languages can vary. Languages can be accusative (which means they attach the same case marker to subjects of transitive and intransitive verbs and a different case marker to objects of transitive verbs) or ergative (which means they attach the same case marker to subjects of intransitive and objects of transitive verbs). Basque is an ergative language:

7a. otsoa   etorri da.
    wolf-DET arrived is (DET = determiner).
    "The wolf has arrived."
 b. ehiztariak   otsoa   harrapatu du.
    hunter-DET-ERG wolf-DET caught   has.
    "The hunter has caught the wolf."

The Absolutive case (the one that includes intransitive subjects and transitive objects) has a zero case marker; the Ergative case (for transitive subjects) is marked by *-ak*. In other words, ergative languages rank thematic role above grammatical function.

In languages like Spanish that have no case marking, an incipient ergativity could be arising in the following way. Because Spanish, unlike English, does not require an overt noun phrase to precede the verb, subjects of intransitive verbs can either precede or follow them (i.e., occur where a transitive-verb object would appear), depending on the relative novelty or importance of subject and verb. (Italics here indicate a shift in rather than a high degree of emphasis.)

8a. Juan ha llegado.
   "John has *arrived*."
 b. Ha llegado Juan.
   "*John* has arrived."

But this is precisely the kind of development that can lead, over time, from a language with SVO order to one with VSO order. Since nearly half verb-initial languages have developed case marking, which is three times commoner among them than among SVO languages, it could be that, at some future date, case markers could develop in Spanish along ergative rather than accusative lines. In other words, it is difficult, perhaps impossible to change any linguistic feature without triggering other, sometimes apparently quite different types of change.

The advantages of grammaticizing verb-argument relations are not limited to improved processing of regular, canonically ordered structures. There is a permanent tension in language between uniformity and liberty in the ordering of constituents. Uniformity is easier to process automatically, but language users have other needs too. As (8) suggests, they seek means to present information in ways that will fit what the speaker knows (or ought to know or should be taking more careful note of). All languages show the results of this tension; all vary in the degree to which they permit flexibility in word order. What most limits flexibility is the hearer's need to know what verb any given phrase is an argument of and what thematic role a particular phrase plays. If it is easy for the hearer to reconstruct this information (in other words, if a language has a sufficiently rich and explicit system for marking these functions), then greater freedom is permitted. A truly free word-order language has proved as elusive as a really free lunch—no language mixes constituents from different clauses, *pace* Evans and Levinson (2009a). Jiwarli (Austin 2001; Austin and Bresnan 1996) comes very close yet lacks the verbal agreement-marking claimed by Jelinek (1984) and others to be what licenses free word order. However, it does have both case marking and head-agreement markers on every modifier of a noun head, so that even where heads and modifiers are separated by other material, constituency is immediately recoverable. Indeed overall recoverability of information seems to be the guiding principle rather than any rigid linking of conditions with prerequisites. In other words, there appears

Variation and Change 173

to be a smorgasbord of relationship-indicating devices, from which languages seem free to draw any selection that meets some yet-to-be-specified criterion of processability.

Grammaticization of Phrase and Clause Boundaries

The third type of response to the limitations of UG stems from the fact that UG does not specifically mark the beginnings or ends of structural units. Take a sentence such as *John found the money Mary had withdrawn*. Were it not for its distinctive intonation contour, nothing would distinguish this from *John found the money. Mary had withdrawn*. It is not enough to say that intonation and context between them would always disambiguate ambiguous structures. Language is not always used under optimal conditions. The environment could be noisy, the hearers distant, distracted, or inattentive. Hence redundancy in linguistic devices is ubiquitous, probably (again) more than nature needed.

Unit boundaries are marked by particular word classes that are final attachments to phrases and clauses. In other words, members of these classes are the last to be attached to their clause or phrase. Here there is a good deal of overlap with the functions of items already discussed. For instance, case markers may serve as right-hand boundary markers. Other boundary markers have only that function, for example, determiners. The set of determiners does not include numbers, although numbers are often final attachments to phrases, because determiners can be attached to their left: [*her* [*three boys*]], [*those* [*five trees*]]. But no two determiners can occur in the same phrase: *[*the his* [*friends*]] (but *the friends of his* okay), *[*my that* [*book*]] (but *my book is that one* okay), and so on. While determiners may also encode semantic information, such as whether a referent is assumed known to the speaker ("definite") or not, there are other ways of expressing this, so boundary marking is their primary function.

Left-hand boundaries of clauses are most often marked by complementizers (what are sometimes called "subordinating conjunctions" in nongenerative grammars). Some of these have additional semantic functions, especially those that introduce adjunct clauses (clauses not subcategorized by the head verb, such as *if, because, although,* etc.). *That* is a clear example of an item that bounds finite clauses, just as *to* is a clear example of an item that bounds nonfinite clauses; they have no independent meaning in these positions. It is only to be expected that in many cases determiners (items that bound phrases) serve as the source for complementizers (items that bound clauses).

There are some languages (English is one) for which the fact that all sentences require an explicit subject makes it possible to use subjects as clause-boundary markers and hence makes complementizers less necessary. The need for complementizers is greater in languages like Spanish, where there may be no overt subject, witness (9a) and its translation (9b):

9a. Who do you think (that) he saw?
b. Quien crees *(que) vio?

(A star outside parentheses indicates that the item(s) inside them cannot be omitted.) However, in a different context, English follows the Spanish example in having an obligatory complementizer:

10a. The man *(that) saw him was John.
b. El hombre *(que) le vio era Juan.

In view of (9), the following contrast may seem at first sight puzzling:

11a. Who do you think (*that) saw him?
b. The man *(that) saw him was John.

(A star *inside* parenthesis indicates that the item inside them *must* be omitted.) Surprisingly, while there is a voluminous literature on the phenomenon represented by (11a) (the "*that*-trace effect"), I have not found any substantive discussion of this contrast or of the related contrast between these two contexts and the usual optionality of *that* in both relative-clause and complement-clause environments. Even a paper that explicitly discusses differences between complementation in N- and V-contexts (Pesetsky and Torrego 2006), specifically citing the *that*-trace effect, makes no mention of it.

To summarize, while in Spanish the "purest" complementizer is obligatory throughout, in English it is obligatory in one context, optional in others, and obligatorily omitted in another. Can we make sense of this in light of the claim that the main function of complementizers is to mark the boundaries of major structural units?

Spanish and English differ in that English, with trivial exceptions, does not mark person and number of subject on verbs and must therefore have overt subject pronouns, whereas Spanish does not need (or normally use) these. It follows that in sentences like (9b), two inflected verbs would directly follow one another if it were not for the complementizer. Since in SVO languages subjects mark clause boundaries, the boundary between main and embedded clause is not clearly marked in Spanish. You might think that it could be easily inferred; a second verb must mean a second clause. But this is not always so (see examples (50)–(52) in Chapter 5), and even if it were, at least one interpretation problem would remain. The only context in which verbs can directly follow one another in Spanish is where the second is an infinitive, and in such cases the subject of the first verb must also be the subject of the second. This is not true in (9b). The inferences required to make sense of sentences like (9b) lie well within the processing powers of hearers, but the automaticity of language processing is built in part on minimizing the role of inference. *Que* does just that with respect to the marking of boundaries and their implications; here it signals "Be prepared for different subjects on V1 and V2," thereby functioning in a similar way to switch-reference markers.

The contexts in which *that* can be deleted are precisely the contexts in which final-argument subjects of lower clauses already mark the clause boundary:

12a. The man [Mary saw] was John.
    b. I think [Mary said [the man was John]].

But where the final argument is absent and the subordinate clause modifies a noun head, as in (11b), *that* becomes obligatory. Omitting it here would allow the sequence *the man saw him,* a complete sentence in itself, and leave *was John* as a baffling asyntactic tag (was *him* or *the man* "John"?). In other words, *that*-omission would automatically generate "garden path" sentences.

The garden-path sentence is a well-studied phenomenon known to severely disrupt comprehension (Christianson et al. 2001; Bailey and Ferreira 2003; Lau and Ferreira 2005). Typical examples include *the horse raced past the barn fell* and *while Mary dressed the baby slept in her crib.* In the first, *the horse raced* is not (as it seems) active but a reduced form of the passive *the horse **that was** raced;* in the second *the baby* is the subject of *slept* rather than the object of *dressed*. The existence of such sentences shows that English has no systematic means of avoiding them. But at least it has no systematic means for generating them; as these examples suggest, a variety of sentence types may be involved, and garden-path cases crop up rarely but across the board. However, if deletion of relative-clause subject *that* was to be permitted, every sentence with *that*-deletion would automatically become not just structurally ambiguous but, like (10a), actually biased in favor of an incorrect interpretation.

Finally, we come to *that*-trace cases. These at first sight might seem the hardest to account for, since the two-verb sequence entailed by (11a) obligatorily leaves the clausal boundary unmarked. However, they have to be considered not as isolated cases but in light of the overall distribution of *that*. Whether *that* is one word with several meanings or functions or several words with single meanings or functions is, of course, a nonissue. The point is that its obligatory use in contexts like (11b), the only context in English where its use *is* obligatory, inevitably creates an association with relative clause (an N modifier) rather than complement clause (a V modifier). Thus a sentence like **who do you think that saw him* would simultaneously do at least two dysfunctional things: it would fill the gap left by the extraction of *who* (thus making it harder to identify the extraction site of *who* and therefore to correctly parse the sentence) and leave open the possibility that the sentence contained some kind of relative clause, a red herring that would interfere with the parsing process.

Thus paradoxically, in this one case the omission of an overt clausal boundary marker helps rather than hinders the discovery of clause boundaries by presenting a verb with no immediately obvious subject and forcing the hearer to identify the gap with the *who* already encountered. In short, though there is no way we can predict variation, variation both within and between languages can be explained if we look at how prior choices of structure or word function affect subsequent choices. What we really need are comprehensive comparative studies of languages that will show how choices of which underspecification(s) a given language repairs first and the means it chooses to repair them will affect subsequent change in that language. In this way we might begin to be able to explain not just the *how* but also the *why* of language change.

Causes of Change

The fact that languages are everywhere variable does not, in and of itself, necessitate that languages should change. It is possible to imagine a world in which languages vary from one another just as much as they do in ours, but where there is no change. Languages could have simply made their initial choices and consistently adhered to them.

Is such a world possible? Probably not, at least not if its languages were structured like ours—perhaps not under any conditions. There are several reasons why this should be so. Perhaps most basic is the simple fact of underspecification. If there is anything whose behavior is not specified in UG, there is nothing to prevent it from changing. Next most basic are the inherently unstable factors referred to earlier: phonological instability and word-order variation, either of which may in turn cause other changes. After these come the multitudinous social and cultural factors that impinge on language: shifts in power and/or prestige, self-identification with conservative or progressive language trends, creation of mechanisms that bond in-groups and exclude, indeed often define outsiders, and so on, practically ad infinitum. Conquest of one group by another may enable the now dominant language to affect the language of the conquered to any extent up to and including complete extinction, but at the very least yielding changes more radical than normal wear and tear could bring in a comparable time frame. Withdrawal of a once-dominant language may create a vacuum that hitherto-marginal languages may fill or that may give rise to an altered version of the original target. Of these last two fates, conquest affected English, while target withdrawal affected several languages that emerged in what had been the Roman Empire. More radical still was the fate of those languages that came into existence as a direct result of European colonialism, in many ways the most revealing extralinguistic factor. These, and their implications for the nature of language in general, are dealt with at length in Chapter 8.

To the extent that these more radical change situations reduce children's access to rich and robust data drawn from a preexisting language, children will fall back on what are, in effect, earlier stages of human language. In an extreme, albeit hypothetical case, input might be so reduced that only the pure output of UG remained, causing language to revert to its condition immediately after the speciation that produced modern humans. Syntax would then consist of nothing more than the basic forms described in Chapter 5, which as we saw suffice to produce a language of sorts, albeit one suboptimal for the hearer. In the real world, it is unlikely that this could ever happen. No input likely to occur naturally would be restricted enough. (An experimentally controlled input would be another matter, but ethical reasons prohibit such experiments.) Words have to be learned if there is to be any language at all, and words tend to drag syntactic implications along with them. Learners take what they can and fit it as best as they can into the framework UG provides. If things still fall short, if key items are still missing, they are supplied by adapting whatever is available.

There are many well-worn routes, used at various times in the course of language change and almost always in the course of creolization. Indefinite articles are supplied by the numeral "one," definite articles by demonstratives, [-real] mood by the equivalent of the verb "go," [-completive] aspect by locative verbs, and so on. If grammatical information is lost through the erosion of bound morphemes, free morphemes will be co-opted. If crucial words, words expressing concepts or functions that language apparently cannot do without—things like possession, existence, reflexive expressions, different types of question—do not occur with sufficient frequency to be picked up from input, periphrastic expressions may appear, chosen from a narrow range of ingredients. For instance, reflexives are typically drawn from words for body parts, though even among creoles there is variation in the part chosen. The creole Morisyen chose the French-derived word for "body," *mo lekor,* "myself,' literally "my body," like thirty traditional languages out of a sample of sixty-two in the *World Atlas of Language Structures,* including Japanese and Igbo. Haitian Creole chose the French-derived word for head, *tet-li,* "himself," literally "head-his," like twelve traditional languages in the same *World Atlas* sample. In other words, during episodes of input reduction speakers are driven by the same forces that, at an earlier stage of language development, drove speakers to create important function words from more transparent lexical expressions.

It now becomes possible to ask (if not yet to answer) a number of questions about the nature of change and variation in language that could not have been posed under previous theories. The theory achieves this result through, first, its systematic distinction between what can and cannot vary in syntax; second, its identification of the function of most variable items (repairing the underspecifications of UG); and third, its division of repair functions into the three categories: marking of phrase and clause boundaries, of relations between connected words, and of relations between verbs and their arguments. Questions that can be asked once this framework is set up include but are not limited to the following: Can we establish minimal and maximal numbers for the different types of repair? Is there any internal implicational ordering in the choice of repair strategies, either positive (if X is chosen, Y will also be chosen) or negative (if X is chosen, Y will not be chosen)? If there is such an ordering, does it extend across all three categories, or is ordering category-internal? If not, is the choice of repair strategies wholly determined by contingency—in a word, anarchic? Do languages tend to focus on one particular category; do they repair in all three more or less equally; or are there different paths, and if so, what prompts languages to choose one rather than another? Is it possible to create a typology of variation and change based entirely on a language's ways of dealing with aspects of grammar unspecified by UG? The width and variety of new fields for future research that are opened up is a further indicator of the value as well as the validity of the theory proposed here.

At the same time, some consequences of the present model must be spelled out, and they are not to everyone's taste. For many, the most aversive will be that a large part of the grammar of any language has to be learned inductively, even by children. The structure of UG may radically reduce the hypothesis space that children can entertain. It will also provide certain expectations as well as certain dis-expectations, one or two of which emerge in the acquisition of English (see Chapter 7). However, after that, children are on their own.

Objections to This Model

Even as I write these words I can hear cries of protest from orthodox generativists. Some of these relate to particular issues, others to one very general issue. If I were to answer all possible particular objections, I would need another book. I shall therefore deal first with just one of these before answering the more general one. Hopefully this first one is the most important of the particular objections; certainly it is one that well illustrates the ways the present model differs from most if not all previous ones.

A Particular Objection: Functional Categories

Readers will probably have noticed that throughout this chapter I have treated things like prepositions, complementizers, and determiners not as major word classes in their own right but as modifiers of either nouns or verbs. This runs counter to virtually all generative treatments of these constituents. Mainstream generative grammar treats all types of word in the same way. Classes other than verbs and nouns are "functional categories" and with regard to structure are treated exactly like nouns and verbs. This means, among other things, that they must be heads. Because they are heads they must have what all heads (at least potentially) have: specifiers (just as nouns have articles and verbs have auxiliaries) and complements (just as nouns have adjectives and relative clauses and verbs have objects, etc.). For example, noun phrases are complements of determiners [*the* [*black dog*]], and the resultant Determiner Phrase may then become the complement of a Prepositional Phrase ([*of* [*the black dog*]]).

Generativists seem quite undeterred by the fact that, in a minimalist analysis, categories like "specifier" and "complement" have no business to be there. Chomsky (1995) claims that there should be nothing in the derivation of a sentence beyond the numeration (the list of words that are going to be incorporated in the sentence, plus, of course, all the properties those words have) and the process Merge, plus anything that can be deduced from those ingredients. This or some similar set of assumptions fits a theory driven by "virtual conceptual necessity," the belief that a theory of language should contain nothing but the absolute minimum required for the production of language. In other words, it makes explicit the kind of assumption that has informed all the best science since the days of William of Occam, if not before.

That every word should constitute a head and that every head should come equipped with a specifier and a complement (or at least with places where such things can be put and hence places that other things, even though heads in their own right, can move to if those places happen to be vacant) cannot be deduced from the ingredients listed either by Chomsky or by the present theory. If these properties and relationships cannot be deduced from the numeration and Merge (or Attach, Close, and phrase and clause algorithms), then they are independently motivated entities that are not conceptually necessary. They should therefore be excluded from any truly Minimalist theory.

The reason for including them in allegedly Minimalist treatments reveals a persistent weakness that runs through generative history: notions developed in one phase of that history are retained in subsequent phases even when changes in the theory render those notions unnecessary or even pernicious. The trinity of specifier-head-complement developed when the sine qua non of syntax was X-bar theory (Chomsky 1972; Jackendoff 1977). Born of a natural desire for uniformity and systematicity in syntactic operations, X-bar theory stipulated that the basic syntactic unit was the phrase, that all word classes were potentially heads of phrases, and that all phrases were hierarchically structured in exactly three levels each: a phrase head XP (X = any word class, and P = Phrase) branched into Spec(ifier) and X' (pronounced X-bar) and X' branched into Head and Comp(lement). As noted in Chapter 2, X-bar theory was explicitly banished from at least the strong version of Minimalism (Chomsky 2007: 4–5), but its stipulations apparently remain unmodified in most Minimalist treatments.

Prior to X-bar theory, most syntacticians took a commonsense approach and treated minor categories as modifiers of major categories. The only reasons for changing the traditional view appear to be the purely theory-internal considerations described here rather than empirical ones. Moreover these are theory-internal considerations that no longer form an integral part of current theory. In accordance with these facts, the present model makes no apology for reducing lexical categories to two main classes: Heads (subdivided into nouns and verbs) and Modifiers (subdivided where necessary into all the other word classes).

A General Objection: Parameter Theory

A common generativist reaction to the overall treatment of change and variation in this chapter, and specifically to the contention that all variable items have to be learned, is likely to go as follows: Surely this is a retrograde step! We had the whole field of variation and change covered with a theory of parameters and parameter setting, giving, in addition to rich linguistic descriptions, by far the fullest explanation of how children acquire language. Why reject the work of literally hundreds of well-trained scholars who have developed this most explanatory of theories?

Part of the answer is surely implicit in the treatment of parameters in Chapter 2. There are simply too many parameters, and the relations between them are too multifarious and lawless for the notion to serve any useful purpose. To say that "the child sets a parameter" has no more content than saying "the child learns a rule." Moreover there are problems with the act of parameter setting itself.

Take the verb-raising parameter (Emonds 1985; Pollock 1989). One of the contrasts between French and English lies in the placement of adverbs (and some other items, such as negative markers) with respect to the verb:

13a. John often kisses Mary.
 b. *John kisses often Mary.
 c. Jean embrasse souvent Marie.
 d. *Jean souvent embrasse Marie.

It is widely assumed that in all languages verbs can assign properties (things like thematic roles and case) to their direct objects only by direct juxtaposition. It follows that all languages must start where English finishes, with the verb immediately adjacent to its object. If that is so, it must be that *embrasse* starts out where *kisses* is in (13a)—that is, the position *embrasse* has in (13d)—and then moves to its left ("raises," since leftward movement is movement *up* a hierarchical tree structure) to occupy the position it holds in (13c). Parameters must be set, as everyone agrees, on the basis of the input the child receives. But the only way a child can know that a verb had been raised is by observing the relative positions of verbs with respect to adverbs, negatives, and so on. So why is "the child sets the verb-raising parameter in French positively" any more than a highfalutin way of saying "the child notices that adverbs, negatives, and so on directly follow verbs in French"?

We haven't even begun to consider the evolutionary problems that parameters face. Baker (2001, 2003) looks at parameters in an evolutionary context more closely than any other generativist, but his conclusions are quite equivocal. In Baker (2001) the possibility that parameters came into existence to serve as boundary markers for potentially hostile groups is dismissed, considering the wide diversity that can exist between neighboring languages, as "overkill"—a single speech sound, as in the biblical *shibboleth,* or the typical Canadian pronunciation of *about,* would have sufficed. Yet group-identity marking plus out-group exclusion is exactly the solution adopted in Baker (2003).

In fact there has never been a valid evolutionary explanation for parameters. If they are to be useful in children's language acquisition without having to be learned, they must be part of UG. But if they are part of UG they must have spread throughout the species. How could differences that distinguish one group from another have come to be the property of everyone? For that matter, one can see why the genome might have spelled out linguistic invariants, but why should it provide recipes for variation, and how could such recipes have entered the genome by any means? Questions like these remain not just unanswered but virtually unasked.

Finally, both some recent work by generativists and some explicit conclusions reached by at least one generativist further devalue the parameter concept. The recent work involves what are known as "microparameters" (Kayne 1996, 2005; Vangsnes 2005; Son and Svenovius 2008). A microparameter is a point of grammar that differs within a group of closely related languages (e.g., the Romance family, Scandinavian languages) or even dialects of the same language, and usually concerns only some minor grammatical feature. Examples in English include the licensing of subject-auxiliary inversion in Belfast English after *wonder (I wonder could he be buying more beer)* and the use of the *for-to* construction to introduce infinitives in some varieties of American English spoken in Arkansas and Oklahoma *(He's gone for to buy more beer).* The discovery of microparameters increases by orders of magnitude the number of parameters that must exist, and to a similar extent reduces the plausibility of supposing that a set of parameters to be fixed by experience forms part of the biological equipment of every human infant.

But if parameters, micro- and macro-, are so numerous and so diverse that they must be excluded from UG, and so intermingled that there is not anywhere, contra original hopes, a small group of master parameters the setting of which would simultaneously fix a whole range of other parameters, parameter settings reduce to a set of facts about particular languages that their acquirers must simply learn in the same way that they learn vocabulary items—by good old-fashioned induction. As Hornstein (2009: 165) points out, "If parameters are stated in the lexicon (the current view), then parametric differences reduce to whether a given language contains a certain lexical item or not. . . . [Parameter setting] is no different from a rule-based approach."

Hornstein's (2009) position suggests that at least some mainstream generativists are moving in a direction similar to that of this book. He too proposes a radically reduced UG (albeit along lines different from those pursued here) and wonders "whether the PLD [Primary Linguistic Data] is sufficient for the LAD [Language Acquisition Device] to construct a grammar given just the invariant basic operations and principles for constructing them" (167). This is a question any theory of an only partly innate syntax has to answer positively if it is to be accepted.

CHAPTER 7

Language "Acquisition"

There is one thing (perhaps the only thing) on which virtually all writers on child language are agreed: that in the beginning there is a child, A, and a language, B, that A does not yet have, and that A aims (consciously or otherwise) at acquiring B by a specifically targeted effort of some kind. This effort may or may not be helped by unconscious knowledge internal to A, depending on one's theoretical bias, but it is one that may best be characterized by the terms *learning* or *acquisition* and treated as a task rather than a set of automatic reactions.

Granted Chomsky has often spoken of the "growth" of language, comparing it to the growth of physical organs such as arms and legs: "Language development really ought to be called language growth because the language organ grows like any other body organ" (Chomsky 1983), and accordingly "language learning is not something that the child does; it is something that happens to the child" (Chomsky 1988: 34). Yet over the years he has shown considerable ambivalence about the nature of acquisition. At the same time as he was making such statements, he was elaborating the notion of a Language Acquisition Device (LAD) that required assumptions no different from those of other models.

The relationship of LAD to UG and the "knowledge of language" discussed in Chapter 2 rather resembles the relationship of the Trinity in Christian theology; they sometimes appear to be distinct from one another but are really aspects of a single entity. The LAD as a repository of "knowledge of language" supplies the child with a set of hypotheses about all possible languages that can be tested against primary linguistic data, enabling the child to identify the grammatical system of the local language (Chomsky 1965, 1972). In introducing the principles-and-parameters model of UG, Chomsky states that as regards "theory of language acquisition, we assume that the child *approaches the task equipped with UG*" (Chomsky 1981: 8, emphasis added). "Growth" can hardly be described as "a task"—certainly not one requiring specialized equipment.

The LAD contained, among other things, a set of parameters, the values of which the child had to set, ideally by simply choosing between two preordained settings. Hence "an explanatory theory [of language acquisition] ought to specify how the learner sets the parameters . . . on the basis of relevant input data" (Dresher 1999: 27). This was normally conceived of as a single, serial process; the child would zero in, as it were, on the correct grammar by a progressive elimination of possible alternatives. However, a different view is equally compatible: if all grammars are potentially available to the child (as they should be if the grammars of six-thousand-odd languages are to be equally accessible and as has been implicitly assumed throughout the history of LAD), two or more grammars could be entertained simultaneously. Each would compete, acquiring points wherever it coincided with primary data, until one or the other won out, and children finally decided they were acquiring English rather than Chinese, or vice versa (Yang 2002).

The seemingly commonsense notion that a language is something "out there" and constitutes some kind of "learning task" for the child is expressed even more forcefully by those who reject the notion of any form of task-dedicated innate assistance. At one end of the empiricist spectrum are those who hold some sort of "general nativist" position—there may be innate processes, but these, unlike the LAD, are not specifically dedicated to language—and who believe that acquisition remains "a process of hypothesis formation and testing" (O'Grady 1987: 173). At the other end are those who may accept that children have "underlying linguistic representations" but claim that these consist of "concrete item-based schemas" rather than "more abstract linguistic 'rules'" (Tomasello 2001: 67) and thus have to be learned in the same way that any other kind of skill or knowledge has to be learned: by induction from primary data, helped by generalization and analogical reasoning.

Even where language is regarded as some form of Platonic idea, discovered rather than evolved or invented, it is apparently not enough for the child to simply rediscover it and put it in motion. "The notion of language as an abstract object does not change the idea of how it might be learned, from the current formal architecture. In either case, hypothesis formation and verification is a possible model" (Bever 2009: 231). The Minimalist Program too might have been expected to make some changes in the generative approach to acquisition. Surprisingly it has not (Longa and Lorenzo 2008). Like most issues that minimalists have not yet satisfactorily dealt with, acquisition is typically offloaded onto the "interfaces"—the requirements of the sensory-motor (read, phonological) and the conceptual-intentional (read, semantic) systems, between which syntax is now seen as no more than a linking mechanism: "Issues of language acquisition . . . also now *presumably* reduce to interface matters" (Larson et al. 2010: 4, emphasis added). Exactly how the issues are "reduced," and what the child actually does in such cases—how its behavior differs, if at all, from that envisaged in previous models—remains unexplored.

An Alternative to the Consensus

What could we say if we took seriously Chomsky's notion that language just grows in the child? Such a notion is clearly in the spirit of the present account. I have claimed here that the core of language is a small set, with perhaps no more than two or three members, of algorithms that automatically create basic structures and that are invariant across languages. Such a mechanism is quite sufficient to generate a full human language, albeit one that would be stripped-down and skeletal in nature compared with any actually existing language. Since the mechanism is invariant rather than aspiring to be a recipe for constructing any of the world's multitudinous languages, it does not require the child to perform any kind of cognitive operation on it or with it. In fact the child behaves exactly as the spider, the beaver, or the bat does; the biological program for what the species does best simply sets itself in motion when stimulated by the words around it.

At the beginning, children learn words. They do not attempt to determine the overall nature of the input they receive. They can have no notion that what they are learning is English, or Yoruba, or Japanese. For the moment, all they are doing is acquiring words that they will then use to communicate with those around them. Once they have found two words that they can combine in order to say something they want to say, they will combine them. They have the innate algorithms—there is no need to hypothesize any kind of developmental delay—but they cannot yet deploy them, simply because they don't yet know enough words to fill the requirements of the algorithms. Take the algorithm for forming clauses: combine a verb and all its subcategorized arguments. Verbs are harder to acquire than nouns, and learning a verb is no guarantee that one will immediately learn the nouns that would combine with that verb to form something the child might have a reason to say. It will be a matter of months before children can fill the framework of even the simplest (single-argument) argument structure with appropriate lexical items. That does not deter them. They do not wait until they have determined how their target language handles a particular structure. Why should they? They have a language already. Their only problem is, though the language they have shares much with the language they will eventually speak "natively," it is not the same as that language. But in the early stages they do not allow this to bother them. They probably don't, at first, even notice any mismatches.

However, as they mature, as they realize that members of any social species have to follow the norms of others, they do start to notice them, as they would notice any other social demands that they are expected to comply with. Only then do children begin to self-correct. But they do not do so on any general or principled basis. They correct one surface form by substituting another surface form. They do this at first on a purely case-by-case basis. It will be some months more before they start generalizing, for example (if their target is English), by following the rule that forms past tenses by affixing *-ed* to verbs and plurals by affixing *-s* to nouns. This procedure—following the innate recipe, adjusting it to fit the primary data when obliged to do so, then storing the result for subsequent use—continues until the stored revisions to the original program amount to a complete grammar of (some variety of) the target language, whatever that is.

Subsequent sections of this chapter will go through the cycle of language development in the child and examine how the claims of this alternative approach affect each stage of that cycle. But before discussing particular stages, a more general issue should be dealt with: whether early child language may legitimately be regarded as a form of protolanguage.

Is Child Language Protolanguage?

One widely criticized claim of the model presented in Bickerton (1990b, 1995) was that the language of young children could be regarded as a form of protolanguage. The most thorough criticism of such claims came from Slobin (2004) and Mufwene (2008); hopefully responding to them will clarify some issues on which earlier treatments may have been insufficiently explicit and even, in one respect, misguided.

Mufwene (2008: 272) does not find "any conceivable parallels between, on the one hand, the early hominids' brains and minds that produced the proto-languages posited by Bickerton (1990b, 2000) and Givon (1998) and, on the other, those of both the modern adults who produced (incipient) pidgins and the modern children who produce child language." His observation is correct but irrelevant. The term *protolanguage* deliberately abstracts away from the natures of its various producers' brains and minds, which of course do differ widely. It does so in order to provide a purely formal description of a system (or perhaps "lack of system" would be a better descriptor) that characterizes a medium distinct on the one hand from animal communication and on the other from true language, yet for those reasons plausible as an intermediate stage between the two. It has no serious competitors for the latter position, so far as I am aware, although unless one wants to countenance a direct leap from an alingual state to modern language in all its glory, some such intermediary must have existed.

Moreover despite differences among its producers, one important reason for the existence and nature of protolanguage is shared by all of them: shortage of words. Without an adequate vocabulary, nothing approaching the syntax of natural language can be developed. Proto-humans did not have enough words because they hadn't yet invented enough words; trained apes do not have enough words because their trainers haven't taught them enough words; pidgin speakers do not have enough words because the social circumstances under which pidgins develop do not enable them to learn enough words; young children do not have enough words because they haven't yet had time to learn enough words.

Grant that in the first two cases there are additional reasons that make true language impossible. Proto-humans didn't have the time (hundreds of thousands of years, at the very least) needed to develop a neural infrastructure for syntax, and apes probably never will. Conversely, pidgin speakers and children do have such an infrastructure in place; pidgin speakers already have at least one full language, and children soon will. But this does not mean that there is necessarily any difference in the forms their utterances take. Indeed in a majority of cases there is nothing in the structure of those utterances that in a double-blind test would show whether they were produced by an ape, an adult pidgin speaker, or a child. That this circumstance should obtain despite the vast mental differences that Mufwene (2008) and Slobin (2005) point out is remarkable, and surely a significant datum in itself.

Slobin (2005) also disagrees on more particular grounds and at a formal level. To him, features of child language are, even at early stages, too rich and too complex to fit the definition of a protolanguage. He points out that the utterances of children learning an agglutinative language such as Turkish or a polysynthetic language such as Inuktitut will differ markedly from those of children learning an isolating language such as English: "Children under 2 who are exposed to such languages do not exhibit the sort of 'pre-grammatical' speech described by Bickerton, Givón, and others, such as absence of grammatical morphology and reliance on topic-comment word order. Turkish toddlers show productive use of case inflections on nouns as early as 15 months of age—that is, productive morphology at the one-word stage" (257).

The typology of languages being as it is, it would be remarkable if things were otherwise. English words typically appear in their stem (citation) forms—unsurprisingly, since there are fewer than a dozen bound grammatical morphemes in the language. Words in languages like Turkish or Inuktitut seldom if ever appear in their bare stem form; in a vast majority if not all cases, they are accompanied by inflections of some sort. Take the following sentence, cited by Slobin (2005):

1. *kazağ-ım-ı at-tı-m.*
   sweater-my-ACCUSATIVE throw-PAST-1ST PERSON.
   "I threw my sweater."

(1) was produced by a Turkish child of eighteen months. An English child of a similar age would probably have said *throw sweater*. But it would have been beyond the powers even of an infant Chomsky to produce *kazak*

atmak—the nearest possible Turkish equivalent to *throw sweater*—since the child will seldom if ever have heard such forms. In other words, children learning a polysynthetic or agglutinative language have no option but to produce multimorphemic words virtually from the beginning, because these are the only words they hear frequently enough to learn.

Note that Slobin (2005) does not claim that Turkish children initially or always use polymorphemic words as correctly as they are used in (1). McWhinney (1976: 398), writing on the acquisition of Hungarian, an agglutinative language that has many structural similarities to Turkish, notes that "word segmentation errors are quite rare for Hungarian children" but that "the segmentation errors that do occur involve separation of one morpheme from another within a word." McWhinney further notes that such errors often reveal incomplete semantic analyses of compound words. In other words, Hungarian-speaking and Turkish-speaking children are simply "learning words" in the same way that English-speaking children do. Words do not come with labels on them saying whether or not they are polymorphemic, so children cannot know this in advance. They cannot analyze words into their component parts, or even know that words have component parts, until they have encountered contrasts such as "sweater-my-ACCUSATIVE" versus "sweater-my-NOMINATIVE" or "sweater-your-ACCUSATIVE." Consequently (1) and sentences like it do not, in and of themselves, provide adequate evidence to demonstrate full mastery and understanding of the inflected words they contain.

Despite this, I am now convinced that child language, unlike early-stage pidgin (sometimes referred to as "jargon" to distinguish it from established and already regularizing and complexifying pidgins), is not really a form of protolanguage. Adopting such a position commits one to the view that the innate component is somehow not available during the first two years or so of a child's life. Then, due to maturation or to the removal of some kind of developmental bottleneck, syntax comes on line, so to speak, and the child transitions smoothly from protolanguage to language.

What's wrong with this position? The apparent existence of some kind of bottleneck operative until the child's third year has been noted frequently in the literature (Bloom 1970; Brown 1973; Pinker 1984; Crain and Lillo-Martin 1999), although there are relatively few attempts to explain it. That the innate component matures over time has been suggested on various grounds by several authors, including Radford (1986, 1990) and Wexler (1998; Borer and Wexler 1987). However, the maturation hypothesis finds itself in opposition to a more widely held position: the continuity hypothesis (Hyams 1986; Boser et al. 1992; Lust 1986).

The continuity hypothesis sees the innate component as constituting the child's initial state; in other words, all of UG is potentially available to the child from birth (Lust 1999: 118). Perhaps the strongest argument in favor of continuity theory is that "there is nothing within the theory of UG to explain how and why UG should be so fractionated [as the maturational hypothesis claims] . . . or why the parts are ordered as they are" (125). The strongest argument against the continuity hypothesis is that if it holds, "the child has a fixed set of linguistic abilities, together with a developing language. The only way to explain this conjunction is to assume that new pieces of language are learned. . . . The maturation hypothesis provides for a stronger theory of innateness than the continuity hypothesis" (Borer and Wexler 1987: 125).

The Borer and Wexler (1987) argument is a powerful one if used against standard interpretations of UG, which contend that all of the grammar of any language (except for a never adequately defined "periphery" of idiosyncratic exceptions) can be derived from UG. Indeed the maturational hypothesis seems at first sight to be more consistent with the empirical data than the continuity hypothesis, which has to claim that those developmental delays that gave rise to the maturational hypothesis are in fact due to the complexity of the child's task in mapping between UG and the raw language data the child receives. This mapping, according to Lust (1999: 142), "takes time." Time, surely—but several years?

As the foregoing suggests, both the maturation hypothesis and the continuity hypothesis have heavy strikes against them, *on a standard view of UG*. However, the words of Borer and Wexler (1987) cited earlier inadvertently show how well a continuity hypothesis fits the view of UG as a partial or skeletal grammar. If syntax is part innate and part learned, the fact that the maturation hypothesis "provides a stronger theory of innateness" than the continuity theory counts against the maturational theory rather than for it. The purpose of hypotheses about language acquisition should not be to provide stronger theories of innateness (or anything else) but rather to provide theories about UG and acquisition that yield a better fit between these and all we know about neurobiology and biological evolution, while remaining consistent with the observed course of language development in children. What Borer and Wexler see as a vice, the present approach sees as a virtue. On this approach, the child has "a fixed set of linguistic abilities" and at the same time "a developing language." Consequently, just as Borer and Wexler claim, "new pieces of language" (those that don't fall out automatically from the fixed set of abilities) do indeed have to be "learned," just as the present approach assumes.

Moreover that approach has at least one advantage over the others. Maturational theory can account for developmental delays only by hypothesizing delays in the maturation of particular components of syntax. But the only motivation for such hypotheses is the delays in learning themselves—a piece of circular reasoning. Continuity theory did provide a reason for those delays (the long time necessarily taken for a child, even aided by UG, to map from primary data to a grammar of the target language), but that account was not particularly plausible and provided no extrinsic reason for why the process should take so long or why its stages should take the course that they invariably do. The real explanation for delays lies in the nature of the words acquired, the number of words acquired, and the order in which they are acquired. In the sections that follow, as each developmental stage is reviewed, readers will see how these factors and the interactions between them play out over the course of acquisition and how they explain the course that acquisition takes over the first few years of a child's life.

As we turn to the early stages, we will see that infant speech does *look* exactly like protolanguage. But prehumans and apes don't have any inbuilt grammar, and pidgin speakers, being adults, have one but find it of little use, since what they are now forced to acquire doesn't have any grammar, and (since they are no longer children) they no longer have free access to mechanisms that would have enabled them to turn structureless input into structured output. Children, in contrast, do have free access to those mechanisms and are merely waiting until they have enough words to put the mechanisms to work.

The One-Word Stage

Recall that the basic assumptions made here are as follows. The child does not "learn" or "acquire" language but rather produces it, just as bats do not "learn" or "acquire" echolocation, and beavers do not "learn" or "acquire" dam building. Beavers produce dams as an automatic reaction to the sound of a running stream (Richard 1983; but see Zurowski 1992). Children produce language as an automatic reaction to the sound of a running stream of speech, with which they are almost constantly bombarded from birth onward (and even before). They learn words because words are essential to their future lives as members of a social, speaking species. They put those words together because they are programmed to do so by the processes discussed in Chapter 5.

I shall ignore the very earliest stage of language development, babbling, since this involves only phonology. I shall also ignore a stage observed by Peters (1983) but ignored in most accounts in which the child seems to be attempting sentences; utterances have sentential prosody but few if any recognizable words. This second phase, however, deserves much more study to determine whether the child is merely imitating caregiver models or at least some such utterances are original. If the latter, this would provide further evidence for the present "coming, ready or not" hypothesis. Consequently, for the purposes of this book, the first developmental stage is the "one-word" stage.

All languages show a period in which words are produced in isolation, although often such words will be followed after a brief interval by others that are semantically or pragmatically related (Scollon 1976). At first glance this might seem to indicate that some form of syntactic bottleneck is indeed responsible. However, Carranza et al. (1991), following spectrographic analyses by Branigan (1979) of both successive and concatenated words, suggest that the avoidance of phonological complications arising when two words are concatenated is a plausible motive.

A factor seldom taken into consideration is sheer shortage of vocabulary. Early words are few in number and very gradually acquired over a period of several months. The child's early vocabulary is such that in fact very few of its words *could* be joined to form any kind of utterance that makes sense. First words show a high degree of uniformity across languages. In most languages, nouns heavily outnumber all other word classes (Braine 1976; Gentner 1982). Moreover nouns of certain types predominate: proper names and common nouns with concrete referents. There typically include "food *(juice, cookies)*, body parts *(eyes, nose)*, clothing *(diaper, sock)*, vehicles *(car, boat)*, household items *(bottle, light)*, animals *(dog, kitty)* and people *(dada, baby)*" (Pinker 1995: 142). Perhaps the next most common category consists not of any true word class but of names for frequently repeated routines and events: *allgone, peekaboo, pee-pee, bye, oh-oh, oops*. (Note that several of these are exclamations that cannot integrate into syntactic structures for speakers of any age.) Verbs and adjectives are extremely rare. The speech of one boy (Ted) reported by Gentner (1982) had no true verbs in the first sixty words learned (over a period of ten months); the most verb-like objects were a past participle, *stuck*, and prepositions like *up* and *down*, which probably functioned as shorthand for *(pick me) up* and *(put me) down*. There were only two adjectives, *happy* and *hot*, but there were forty-one nouns.

As the foregoing suggests, most of the theoretically possible combinations from such lists (e.g., *sock juice, happy nose, oops peekaboo*, etc.) would be either meaningless in themselves or nothing that a child under two might reasonably be expected to want to say. In the absence of any clearly definable motivation for one-word or two-word limits, the null hypothesis should be that as soon as children learn any words they can meaningfully put together, they put them together. In concrete terms, if Gentner's (1982) subject Ted never said *sock up* or *up sock* when one of his socks fell to the floor, that was more likely because he had no interest in rescuing his sock than because of any incapacity to concatenate *sock* and *up*. It should be relatively easy to test this hypothesis by dropping socks to the floor (or performing any other action that might trigger the conjoined utterance of any pair of words the child knows) and then observing the child's reaction, verbal or otherwise. But so far as I know, no one has yet done this.

At least three factors combine to determine what words are learned first: ostensive reference, frequency, and phonological salience. Nouns with ostensive reference are those whose meaning can be learned from the physical presentation of their referents, with or without benefit of parental explanation, nouns like *book* and *lamp* rather than nouns like *fear* and *mercy*. Such meanings are much easier to grasp than those of verbs, even the most concrete of which—*fly*, say, or *break*—have to generalize across a wide range of fliers and breakages. Among nouns that can be defined ostensively, those that occur most frequently will naturally be the first learned; for instance, among body parts, *foot* was known to all members of a sample of twelve children, *armpit* to none of them (Goldin-Meadow et al. 1976). As for phonological salience, nouns and verbs are more salient than things like prepositions or auxiliaries; they carry heavier stress and don't undergo contractions or elisions.

It should be borne in mind too that the separation of any words at all from the stream of speech is no mean feat. Listening to a recording in a language that one doesn't know is an edifying experience here, because this is exactly how the future language of any child must initially sound to that child. One factor that makes it possible to divide speech into words is statistical learning (Saffran et al. 1996; Kuhl 2010). The child determines word boundaries by the relative likelihood of transitions between sounds: the less frequent a transition is, the likelier it is to occur between rather than within words. Clearly it will be easier to compute boundaries for words that always take the same form and that never undergo the kinds of change associated with contractions and so on that are typical of grammatical as opposed to lexical (referential) items.

The Two-Word Stage

The one-word stage frequently (though not always) lasts several months. The two-word stage is considerably shorter. Common sense suggests that as the child progressively escapes the restrictions of infancy, a wider range of capacities is deployed, synergies result, and the pace of development consequently quickens. But common sense isn't always (or even often) science, and I know of no cogent proposal involving cognitive development that would explain why the two-word phase should be shorter than the one-word phase, or even why either of them should exist in the first place. Vague appeals to "cognitive limitations" or "developmental phases" are not enough.

Early attitudes to early child language, up to the 1950s, were that the child learns by imitation, analogy, and the like. These beliefs could hardly have survived a more thorough study of the raw data. But before such study could develop, *Syntactic Structures* changed the entire ballgame.

The child was thenceforth seen by many researchers as forming "grammars" of some kind at every stage from two words up. One early attempt at such a grammar (Brown and Fraser 1964) merely listed and sorted combinations, selecting just two "word classes" based on whether words occurred in first or second position in utterances. A rival approach (Braine 1963), "pivot-open" grammar, saw a handful of words like *more, off,* and *allgone* operating as "pivots" (capable of appearing as either first or second word) with a large "open" class consisting of all other words (hence inclusive of nouns and verbs alike). Such attempts, while they might have been observationally adequate (although pivot-open grammars weren't even that), failed either to give any adequate explanation of what the child was doing or to relate the two-word stage to subsequent stages, let alone the final adult form of the grammar.

Recall that the innate algorithms provide no information on the order of adjacent constituents. Words in a phrase or clause may attach to the left or to the right of their heads. Experience is the only guide here, yet early in the comparative study of acquisition across different languages it was observed that children very rarely make mistakes in the order of constituents (Bloom 1970). Note that if one gets pairwise attachment right, there is no need for any further mechanism for determining the overall word order of the language: if a precedes b and b precedes c, then transitivity gives you that a must precede c. In fact once one has determined what the words are, determining the order of words is a far easier task, which must be a contributory factor to the difference in length between the one- and two-word stages.

The two-word stage serves to establish some of the major word-order relationships in the target language: subject-verb, verb-object, adjective-noun, and so on—a vital building block for the operations that develop in subsequent stages. Note that, as with the one-word stage, there is no limit on child production that prevents longer utterances in principle, but development is still delayed by small vocabulary size and conditioned, to a large extent, by the nature of vocabulary growth.

Although some recent work has cast doubt on the reality of a "vocabulary spurt" (Ganger and Brent 2004), there is no question that the pace of word learning picks up as the second birthday approaches and continues at a rapid rate through the next couple of years or more. Numerous explanations have been given for this phenomenon (see Nazzi and

Bertoncini 2003 for a convenient summary of these). However, there is no need to assume that any particular cognitive development is involved, since Zipf's Law (Zipf 1935) adequately accounts for the relevant phenomena.

As we saw in the previous section, frequency is one determinant of what words are learned first. Zipf's Law states that the frequency of any word is inversely proportional to its rank in the frequency table (Zipf 1935). It follows that, ceteris paribus (and things aren't that equal, since salience plays a large role and the commonest words are seldom salient), the commonest words will be learned first. But Zipf's Law means, among other things, that a minority of types will yield a majority of tokens. For instance, though a mature speaker of English may have a vocabulary of over fifty thousand words, the commonest two thousand words yield 80 percent of the text of the million-word Brown Corpus and 96 percent of the text of informal speech (Francis and Zucera 1982).

In consequence, a cycle of lexical development forms. The first fifty to one hundred words (a normal tally for the one-word stage) are acquired very slowly. Even allowing for the fact that comprehension always outstrips production, the text even of motherese (the simplified register used by caregivers in some but not all cultures) must exceed the knowledge of children at this stage; much adult speech probably sounds to the child something like "No blah blah blah blah dog blah blah car blah blah." The child may be able to figure out from context and general pragmatics that mother means "No, you can't take the dog in the car with you." But lack of salience prevents the child from getting the very commonest words. The function words (grammatical items)—pronouns, prepositions, and the like—may act as useful structural signals for the older child but from a younger child's perspective are so much useless padding. So the slow pace of early word acquisition is caused in part by the fact that the child has to pull isolated plums of referential and ostensively definable sense out of a pudding of mysterious ingredients.

Another factor is what you might call "lexical isolation." First words represent a wide scattering of meanings that don't necessarily have much connection with one another. But having a workable vocabulary entails building a word web, a collection of words partly definable in terms of their relations with one another: a dachshund is a dog, a dog is an animal, an animal is alive and a hammer isn't; red is what's left when you reach the boundaries of brown, orange, and pink; and so on. The existence of such a semantic network is treated by some writers (e.g., Deacon 1997) as a prerequisite for any kind of truly symbolic system. Surely it is an aid in word learning: the denser the network of meanings you build, the fewer the holes left in it and the easier to fill those holes by learning the appropriate word(s) on fewer and fewer exposures as the word web becomes more complete. This process forms a powerful contributor to the process known as "fast mapping" (Bornstein et al. 1976; Carey and Bartlett 1978; Heibeck and Markman 1987), in which children, usually well past the two-word stage, will learn words on as little as a single exposure. But continued word learning also makes possible the full comprehension of more and more sentences, not the spotty, largely pragmatic comprehension described in the preceding paragraph. And full comprehension in turn makes more words usable in the child's own productions.

In other words, word learning is an autocatalytic process. Entry into the two-word stage is contingent on developing a critical mass of words. Transitioning from the two-word to the next stage is simply a function of the increasing rapidity of vocabulary growth that this mass makes possible, and for that reason the two-word stage is (absent severe developmental problems) much shorter than the one-word stage.

Before we proceed to the next stage, however, it may be worth briefly glancing at one grossly underresearched group. As Nelson (1981: 215) noted, "We have all observed children who go virtually through the second birthday, or even later, without producing language and then, at about 25 months, begin to produce sentences that are roughly equivalent to those of children who have been producing right along." Are such cases exaggerations or even myths, like the alleged first words of Lord Macaulay: "Thank you, madam, the agony is sensibly abated"? If they are not (and it shouldn't overtax the abilities of graduate students to find cases of apparently otherwise normal children of two or more who haven't spoken yet, and study their subsequent development), and if their first sentences are as Nelson suggested, this would have significant consequences for acquisition studies, casting doubt on any theories that include hypothesis testing or the essentiality of communication and adding support to the claim that language deploys automatically as words are learned.

"Telegraphic Speech"

What follows the two-word stage has traditionally been described as "telegraphic speech" (Brown and Fraser 1964), a medium more recently described as consisting of "strings of words . . . in phrases or sentences such as *this shoe all wet, cat drink milk,* and *daddy go bye-bye,*" although at the same time "a number of grammatical inflections begin to appear," and "simple prepositions *(in, on)* are also used" (Yule 2010: 175). As Yule's definition suggests, the "telegraphic speech" stage has no clear boundaries and, like the stages that precede it, is no more than a convenient label for phases of the continuous unfolding of the language faculty.

If the two-word stage was largely concerned with determining the linear relations of word pairs, the telegraphic stage is largely concerned with things like grammatical inflections and simple prepositions. Up until this point, the child acquiring English and similar languages has had to deal only with the word classes UG explicitly deals with: nouns, verbs, and modifiers of these. Paradoxically (and for English speakers, quite counterintuitively) this fact may make languages like English and Chinese harder to acquire than highly inflected languages. In highly inflected languages, children are from the very beginning spoon-fed with bite-size bits of the unscripted parts of their language—that is, aspects of language that emerged post–UG formation and that therefore have to be learned. These things are actually attached to words they have to learn, which gives them a head start of maybe a year or more over children with isolating, inflection-poor languages.

On the other hand, in languages like English and Chinese, most words appear free of those bits, and the bits themselves, brief, unstressed, minimally semantic, and subject to heavy phonological distortion, float in limbo between easily recognizable bare nouns and verbs. No wonder one finds what you might call the *throw-sweater/kazagimi-attim* effect—the fact that children learning a highly inflected language show greater grammatical sophistication at an earlier age than children learning an isolating one like English.

However, even a child learning English must sooner or later deal with a class that, after all, includes the commonest vocabulary items in the language. As vocabulary grows and longer utterances can be more fully comprehended, more and more of these grammatical morphemes are brought to the child's attention. They present two distinct problems. One involves their meanings, many of which are highly language-specific (in contrast to those of referential words, which usually have at least a core of consistency across languages) and elusive in ways that will be examined in a moment. The second problem involves their sounds; they frequently take varying forms, due to a combination of low stress with elisions or contractions that depend on the speech sounds that occur immediately before or after them.

Little of theoretical interest can be said about the second problem. The first is another matter. Consider all inflections, particles, and "function words" as constituting a single class: the class of grammaticizable semantic distinctions. They include, but are not limited to, things like gender, tense, case, number, person, and so on. Is the list finite? Almost certainly it is, and probably quite short too. Some distinctions (e.g., evidentiality markers on verbs) are relatively rare, while some (e.g., plurality on nouns) are very common. The question that arises is whether the category "grammaticizable distinction" is in some sense known to the child, who may even have intuitions about things that could be grammaticizable distinctions and things that could not.

The task of acquisition would be greatly lightened if some precognition existed, but such evidence as there is appears conflicting. In creole languages, as noted in Chapter 8, particular types of tense, aspect, and modality appear with a frequency well beyond chance, even though the system constituted by these types differs from those found in most if not all of the creole's antecedent languages (Bickerton 1981). However, hardly any creole has simple past with moment-of-speech as its reference point, which does occur in English child speech, even when the form chosen to express it is not a correct form in the target language. Seth, the child of one of my former students, seemed at one stage to be actively seeking some grammatical marker for simple past tense, and after experimenting with different forms, for several weeks settled on *didja* (not necessarily implying a question):

2. Didja toot.
   "I (just) farted."

What seems the likeliest possibility is that general cognitive factors supply a list of common distinctions that may (or may not) be grammaticized in any given language. Any item on that list can be acquired easily on the basis of positive evidence, but whenever the child finds no clear evidence for a particular distinction, a default assumption applies.

As with the one- and two-word stages, the so-called telegraphic speech stage does not correspond to anything substantive in language development but merely reflects the growth of vocabulary. There is no corresponding stage in polysynthetic languages. It may be illuminating to consider the significance (and some consequences) of this.

Polysynthetic and isolating languages are, in their purest forms, merely extremes on a bimodal distribution in which many languages mix elements of each type. However, mixtures are seldom even, so most languages are predominantly isolating or predominantly polysynthetic. In polysynthetic languages grammatical items are embedded in words so that the child can, at least in principle, deploy the following discovery procedure. First, factor out the part(s) of a word with lexical content—that is, the part(s) that refer to some entity or activity in the world. What is left must have grammatical function(s). Check against other words and substitutions within them to determine how many separate grammatical items there are in each case. Figure out from context what each one's function is. Provided that the form of a grammatical item does not vary too unpredictably in different contexts, the child's task is not too dauntingly difficult. It is in fact only a more circumscribed variant of the task all children face when confronted by any language: the task of factoring out individual morphemes from the unbroken stream of speech. So this may indeed be easier to do than to locate and identify items that occur, seemingly randomly, between things you already know.

However, two things can be said with relative certainty. The first is that differences in learnability and any consequent pressure on languages to adapt cannot be particularly great; otherwise we would all be speaking something like Turkish or Inuktitut. The second is that the relatively slow acquisition by English-speaking children, and the problems they have in acquiring grammatical items, cannot arise as a result of some form of bottleneck or developmental delay. Nor can it be due to maturation of an innate grammar, since it would be absurd to suppose that Turkish has a faster maturing grammar than English or vice versa. It is the nature of the target language rather than some internally regulated pace of linguistic development that is responsible for different patterns of acquisition.

The contrasting patterns in Turkish and English acquisition lend further support to the proposal that grammar has learned and unlearned components. Suppose that there was a UG that was universal in the sense discussed in Chapter 2, that is, capable of generating the grammars of all the world's languages. If this were the case, there would be no reason for the same grammatical functions to be learned more quickly and easily in some languages than in others or for differences in the ways grammatical relations are expressed to cause different rates of acquisition in different languages. Note too that it is precisely in areas un- or underspecified in the kind of UG described in Chapter 5 that interlinguistic differences in the pace of acquisition occur. If all grammars were equipotential, as the classic UG/LAD entails, such effects should not occur. Children should have no expectations whatsoever about what kind of language they were going to encounter. However, in real life they behave as if they did indeed have expectations, and where those expectations are frustrated, delays in the acquisition of particular structures occur.

The sections that follow look at three areas that cause difficulties for children acquiring English: the causative/noncausative distinction, sentence negation, and the syntax of questions. In all these areas the pace of acquisition is, for children acquiring English, slower than acquisition of the same areas in languages where there are no distortions of UG-predicted patterns, and slower too than in other areas of English grammar where those patterns are followed. Current theories of acquisition have no satisfying explanation for the data surveyed here.

The Causative/Noncausative Distinction

While there exist a few English verbs that can be used both causatively and noncausatively (e.g., *melt: The ice melted* vs. *I melted the ice; feed: The cows are feeding* vs. *She is feeding the cows*), in many cases verbs require an external causative, *make (He made X do Y)*. In others there are pairs of verbs (e.g., *teach, learn*) that are semantic mirror images of one another: one inherently causative and one inherently noncausative. However, many children treat a wide variety of noncausative verbs as if they were causative. Bowerman (1974) gives seventeen examples of noncausatives used as causatives by the same child, and lists one hundred examples from other children, featuring thirty-six different verbs, all produced before the children had reached age four. Examples (3) and (4), for instance, were both produced by Bowerman's daughter before age three:

3. Go me around.
4. Can you stay this open?

This suggests that a rigid causative/noncausative distinction for verbs does not form part of the innate grammar. Why should it? Absence of stipulation is one of the things that characterize the kind of UG envisaged here. In Chapter 5 we saw how absence of any stipulation as to what could or could not constitute a modifier licensed the modification of one verb by another verb and thereby made possible serial verb constructions. Accordingly it seems likely that in earlier stages of human language, all verbs could potentially be used in either causative or noncausative senses. Inevitably in actual use some verbs would be used in one sense much more frequently than the other, leading to restrictions in their meaning that, since they are learnable, would simply be learned.

However, the data for causative/noncausative acquisition in English present a problem for a number of existing claims about how acquisition in general works. Take, for instance, the proposal by Baker (1979) that, given the lack of overt negative evidence (being told explicitly that certain sentences are wrong), children would hypothesize a rule only if that rule could be confirmed on the basis of positive evidence; in other words, they would not take the riskier path of guessing at a rule simply because they had found nothing in their input that would contradict it. This was refined into the "subset principle" (Berwick and Weinberg 1984): children would always hypothesize the most restrictive rule—in this case, that a verb used noncausatively could only be used noncausatively—and maintain that rule unless positive evidence (the same verb used causatively) showed that the rule was too restrictive.

But if the subset principle worked across the board, children would never take a verb they had learned as a noncausative and use it as a causative. Since verbs like *go* and *stay* are never used causatively in adult speech, the subset principle should prevent children from ever producing sentences such as (3) and (4). The existence of verbs like *feed* or *melt*, even if children had happened to encounter such verbs used in both ways, would not constitute evidence for the less restrictive rule. It would merely indicate that some verbs incorporated both properties, leaving the child to discover which verbs behaved in this way on a case-by-case basis. Mistakes made through overextending causativity can never be repaired on positive evidence only.

Evidence from the causative/noncausative distinction is just as damaging to empiricist theories of learning. Recently there has been considerable support for the notion of "item-based learning," which claims that "children's early utterances are organized around concrete and particular words and phrases, not around any system-wide syntactic categories or schemas" (Tomasello 2000a: 156). For instance, while children may be able to produce *the window broke*, they "cannot go on to produce *He broke it* or *It got broken*, even though they are producing simple transitive and passive utterances with other verbs" (Tomasello 2000b: 210).

But if children cannot even generalize to legitimate constructions, how is it that they are able to overgeneralize to illegitimate ones? Tomasello (2000a: 158) himself has to admit that children can learn noncausative verbs and then use them causatively. He tries to avoid this counterevidence by claiming that children "produced very few of these types of overgeneralizations before about 3 years of age" (158) and cites three sources (Bowerman 1982, 1988; Pinker 1989) to support his claim. But his cited sources are general summaries ranging over numerous child language phenomena, and while they mention a few examples across a wide age range, they give no specific information on and make no specific claims about how early transitive use of intransitives commences—although they do show that such errors continue to occur up to age six or even later (examples in Bowerman 1988). This is precisely because, in the absence of parental correction, only the continued absence from primary input of sentences like (3) and (4) over a long period can eventually convince the child to abandon such expressions. Significantly Tomasello fails to cite Bowerman (1974), a paper, unlike those he cites, that deals extensively and exclusively with the causative/noncausative issue. Of seventeen mistakes involving noncausatives used as causatives made by Bowerman's daughter, no fewer than nine appeared before the age of three. Such errors therefore cannot be dismissed as some weird aberration of later learning, and it's not clear what difference it would make even if they could.

Causative/noncausative errors support the view that children are not doing the kinds of thing that acquisitionists, empiricist and nativist alike, claim they are doing. They are not learning by rote the permissible structures for particular verbs; they are not forming hypotheses based only on positive evidence; and they are not restricting their hypotheses to a narrow subset of possible ones. They are forging ahead with their own ideas about what the syntax of the target language should be, and they are surprisingly resistant to counterevidence. Evidence from two further areas supports this view. Both areas are among the most thoroughly studied in the field, with a large body of work going back more than half a century. All these studies, however, approach the problems from either inductive-learning or orthodox generative assumptions, so we should reexamine them too through the lens of the present theory.

French versus English Negation

The problems that English-speaking children have with negative sentences are well-documented. Those problems arise partly because the sentential negator *not* seldom appears alone in spontaneous discourse but typically undergoes vowel reduction to *n't* and subsequent combination with auxiliaries that may themselves appear in reduced forms. Consequently *not* gets lost in a forest of *didn't*s, *don't*s, *won't*s, *hasn't*s, and *aren't*s and is hard for the child to access. Conversely exclamatory and noun-phrase negator *no* is highly salient, occurring frequently in isolation and with relatively heavy stress.

The situation of French-speaking children is quite different. Formal French has two sentential negators, *ne* and *pas,* one preverbal, the other postverbal. However, the first is rarely pronounced in colloquial French, while the second is never reduced or elided. Moreover French is a syllable-timed rather than a stress-timed language, so there is far less difference in salience between *pas* and its accompanying verb than there is between *not* and its verb in English.

These facts play a highly significant role in the development of negation in the two languages. Since English speakers normally acquire *no* in their early vocabulary (and usually long before *not*), expressions like *no ball* almost invariably occur in the two-word stage. It is therefore unsurprising that in the telegraphic stage, before auxiliary verbs are learned, sentence-initial *no* continues to be the sentential negator of choice. Consequently one finds sentences like *No the sun shining* and *No I see truck* (Bellugi 1967). Shortly thereafter sentences such as *He no bite you* and *I no want envelope* begin to appear. Of three children whose output is summarized in Deprez and Pierce (1993), one had reached 50 percent preverbal negation by age 2.4, and one had reached 70 percent by age 2.0, though *no* rather than *not* was oftener used. Significantly, identical forms of negation are found almost universally among creole languages, although they never appear in any variety of adult English.

Is there a canonical position for negative markers in UG? If there is, logically it should be immediately before the predicate, since such markers negate only the content of the predicate, not the existence of the subject. But if that is so, what accounts for the earliest phase of presentential negation? Either the prepredicate preference is not in UG, or *no*+S simply extrapolates from the two-word *no*+N pattern into the start of the telegraphic stage.

Negation develops differently in speakers of French. French-speaking children under the age of two have well-established *pas* in all the right places: before infinitives but after finite verbs. Distinguishing between finites and nonfinites is hard for English-speaking children because the forms of present tense and infinitive are identical except for third-person-singular (e.g., *break, [to] break*). However, it is much easier for French-speaking children, since French stem forms (e.g., *roul-*), can be derived from the occurrence of inflected forms—*roulons, roulez, rouler* ("We roll, you [pl.] roll, to roll")—and the fact that the infinitive itself is inflected. Consequently while nonstandard forms of negation in English persist well into the third year, they are virtually absent from the beginning in French. (Pierce 1989 shows that sentences misplacing *pas* with respect to the verb account for less than 3 percent of her data.)

Do we need any further assumptions to account for the difference between the two developmental patterns? Those who approach the issue from a generative perspective have proposed explanations based on abstract principles and covert processes. According to Klima and Bellugi (1966), English negative-placement errors result from children's failure to acquire a negative-lowering transformation. Deprez and Pierce (1993), on the other hand, hold that English negatives are really in the same position in both child and adult; what happens is that children fail to perform a subject-raising movement that would carry the subject above (hence to the left of) verb and negative. Conversely they attribute the virtually error-free negative production of French children to their early acquisition of verb raising, obligatory for finite verbs (since, under this set of assumptions, verbs must raise to acquire tense) but barred for nonfinite verbs (which are, of course, always tenseless).

What would have to be true if such explanations were correct? One answer would be that the child's brain must somehow represent much more than words plus recipes for assembling phrases and clauses; it must also represent grammatical categories, not just "noun" and "verb," but things like "subject" and "tense." More than that, it must represent hierarchical tree structures with empty spaces in them for things to move to and labels for those places. The fact that no such things are accessible to consciousness is, of course, totally irrelevant. But whether these are plausible mechanisms for brains to have is another matter entirely, and surely a legitimate issue.

A generativist might respond, "Well, we don't really know how the brain represents *anything*." This is no longer true. We know, for instance, that "the representation of a word is neurobiologically realized in mostly cortical networks having a topology that depends on the modalities correlating with external aspects of the meaning" (Assadolahi and Pulvermuller 2001: 311) and that the neuromagnetic signals of individual words can be detected with an accuracy far beyond chance (Kellis et al. 2010). Nothing remotely approaching this level of understanding is available for representations of the abstract categories and structures that would have to exist if the proposed mechanisms for negative acquisition (and countless similar proposals by generativists) were to actually work. Indeed, so far as I know, there are not even any speculative accounts of how and with what the brain might construct the necessary representations.

Of course, none of this means that such representations cannot exist. It is simply an issue of probabilities and consequent research strategies. A normal research strategy is to start with some parsimonious and relatively well-supported proposition and take it as far as one can. If this resolves the issue, no alternative needs to be considered. Only when that process breaks down will the researcher begin to try alternatives. In the case of negative acquisition, an account based on word learning seems perfectly adequate.

Assume that all children know (whether innately or from experience is immaterial here) that a sentence can be negative and that something in the sentence will mark this fact. English-speaking children then seek to identify a negative marker but are unable at first to correctly do so, mainly for phonological reasons. Consequently they begin by choosing the wrong marker *(no)* and putting it in the wrong place (sentence-initially instead of predicate-initially) because, after all, this is where *no* typically appears in adult speech. English-speaking children, failing to locate (or perhaps even identify) any negative marker in the "right" position, do revise what they had first assumed and begin to place *no* predicate-initially, even though there is not a single example of this in their input. They persist with this behavior until they begin to acquire, most likely on a piecemeal basis, a string of reduced negatives—*aren't, don't,* and so on—from which evidence the correct negative placement can eventually be factored out.

Conversely French-speaking children avoid these problems by having from the start a readily identifiable negator *(pas)* and two clearly marked positions for it, immediately before or after the first verb (whether main or auxiliary), either of which gives it full predicate scope. In other words, all they are required to do is to observe and obey the relative positions of adjacent words—that is, set a relationship left unspecified by UG, which determines only dominant-subordinate relations based on order of attachment. As we saw in considering the two-word stage, paying attention to the order of words is a major preoccupation that prevents all but a tiny fraction of ordering errors even in the earliest stages. (It also happens to be one of the six "operating principles" that Slobin 1973 attributes to child learners.) Moreover word order is the only evidence on which children could base the assumption that verbs "raise."

Thus the course of acquisition in both languages can be better explained by the most parsimonious explanation—the child learns words only, has problems learning function words, but forges ahead anyway, using inbuilt schemas—than by assuming any process involving transformations, covert movements, or any other kind of abstract apparatus. This makes it unnecessary for us to hypothesize any otherwise unsupported brain mechanisms that might somehow represent such abstractions. This conclusion is strongly reinforced by data on the acquisition of interrogatives, which illustrate the tenacity with which children can resist the evidence of caregiver speech when such evidence goes against their innate recipes for sentence building.

Acquiring Question Forms

In English a very large proportion of adult utterances to children consists of questions: 33 percent according to one study (Cross 1977), 44 percent according to another (Newport et al. 1977). The inversion of auxiliary and verb that characterizes most of them, WH- (*who, what, where,* etc.) and yes-no questions alike, should therefore be more familiar to children than almost any other feature of English. If children were learning inductively and without any inbuilt expectations, correct question structures should be among the first things they acquire.

However, "questions are perhaps the only syntactic structure for which English-speaking children commonly make word-order errors" (Ambridge et al. 2006: 520). Indeed it takes children two years or more to fully master these structures (Klima and Bellugi 1966). The earliest questions, in the two-word stage, typically consist of a WH-word plus a noun, like *where kitty?* Subsequently questions are produced as declarative structures, their interrogative nature shown only by intonation. This is the structure used in creole languages, as well as in languages like German where questions marked only by intonation form a legitimate alternative to auxiliary- or verb-subject inversion. German children acquire this question pattern early and effortlessly, whereas the subject-verb inversion type is not acquired until much later (Mills 1985). In Kaluli, a language of New Guinea, an invariant interrogative marker for noun phrases was acquired by twenty-five months, but sentential interrogative particles that required a finer person distinction in the verb caused problems (Schiefflin 1985). In Hebrew yes-no questions are formed with declarative order and marked by intonation only, while WH-questions require only the question word placed at the beginning of a declarative structure. According to Eyal (1976), nearly 80 percent of a sample of Hebrew-speaking children under thirty months produced errorless questions, whereas English-speaking children did not reach this level until the middle of their fourth year.

The delay in acquiring question forms in English might be supposed to have something to do with a comparable delay in learning auxiliary forms. These indeed are often contracted and carry low stress, which might make them a poor target for word learning. Delayed auxiliary learning doubtless contributes to the difficulty of acquiring questions. However, it cannot entirely explain it, because in yes-no questions auxiliaries typically appear uncontracted and occasionally even stressed. (***Did*** *you pull the doggy's tail?*) Relevant here is the Auxiliary Clarification Hypothesis (Richards 1990), which proposes that since children are biased to attend preferentially to the beginnings of sentences, auxiliaries are more readily processed and learned more easily when they occur in questions. If this were really the case, we would predict that an increase in the frequency of auxiliaries in caregiver speech would facilitate children's acquisition of correct question forms. However, in a study that both reviewed the literature and conducted experimental tests (Fey and Loeb 2002), the authors found that increased use of auxiliaries in questions by caregivers had no significant effect on children's subsequent production.

Indeed available evidence from a number of different sources seems to consistently support the proposal that, for child learners, there is something bizarre about English question formation. It is as if children come with the expectation that questions should not involve order changes but should be expressed through intonation change, perhaps helped out with an invariant morpheme that signals interrogative mode (as is the case for negatives). When this expectation is violated, and especially when the violation involves some exotic feature such as subject-auxiliary inversion, children are baffled, and it may take years rather than days or months for them to fully acquire the correct forms.

Note that while this proposal fulfills the predictions of the present model, the data to be explained are as puzzling for theories of inductive learning as they are for theories in which the child is guided by UG. Theories of both types are, and necessarily have to be, neutral with respect to the world's languages.

Inductive learning theories cannot assume any kind of built-in preference on the part of the child for one type of syntactic structure over another, because nothing is supposed to be built in. For inductivists, the neonate's brain is a blank slate (apart, of course, from whatever faculties enable inductive learning), and therefore everything has to be derived from experience. But since the majority of children in modern nation-states get exposed to only a single language, no preferential bias would be obtainable. Subject-auxiliary inversion cannot be judged more difficult than other types of structure because there does not exist any objective, universal metric of linguistic difficulty, and subjective metrics are determined by the language one happens to have learned oneself.

To an adult speaker of English, the mechanisms of an agglutinative language like Turkish may seem horrendously difficult—certainly far more difficult than anything in English and surely a harder learning task for the child. Yet according to Aksu-Koc and Slobin (1985: 845), "The inflectional system [of Turkish] appears early, and the entire set of noun inflections and much of the verbal paradigm is mastered by 24 months of age or earlier," an age at which many English-speaking children are still struggling to get past bare two-word utterances. In addition to explaining how such things can be, recall that inductivists also have to explain why English interrogatives should take so long to master when at least a third of the input consists of them.

Standard nativist accounts must be equally language-neutral. UG cannot admit a bias for some languages or language types. All human languages, even all types of structure, should be equally easy to learn, with allowance for varying degrees of complexity and input frequency. Neither factor applies here. Questions have high frequency in caregiver speech, and their complexity is low, requiring only, in the case of yes-no questions, one movement, and in the case of WH-questions, two. Recall that in the case of negatives, generativists routinely assume that covert movement occurs very early. Why should overt movement be a bigger problem than covert movement?

In all three areas, neither empiricist nor nativist proposals give satisfactory results. Yet in all three the evidence is fully congruent with a theory that assumes a limited innate blueprint for syntax and leaves its detailed local execution to be acquired inductively—by trial and error, one might almost say.

"Error" as a Source of Insight

The foregoing suggests a useful way of looking at child "errors." The kinds of error that children typically and repeatedly make should help us to distinguish more precisely between what UG makes available to them and what it does not. The latter are things that have to be learned, and we would therefore expect them to take more time. Cross-linguistic studies of error (and its relative absence in some languages) will further refine this process. If a particular feature is acquired easily in language A but only with considerable difficulty in language B, this suggests two things. First, acquisition where it affects functional categories could be easier in highly inflected than in poorly inflected languages for the reason given earlier: inescapable ingestion of grammatical morphemes at a much earlier age. Second, acquisition of some structural features could be much harder in languages that have sharp and unexpected deviations from the simplest ways of expressing basic patterns laid down by UG.

At this point it is worth beginning to consider the significance of creole languages for acquisition studies. These languages will be central to Chapter 8, but since they have an important bearing on acquisition issues, the present chapter would be incomplete without some reference to them. Bickerton (1981) was the first source to propose a connection. That work noted a number of "errors" in language acquisition that would have been perfectly grammatical in a creole language and suggested the unlikelihood of there being two acquisition devices, one for use when input was robust and uniform, another for use when input was chaotic and deficient. If there was a single mechanism designed not for discriminating among a range of possible languages but for producing a language of a particular type, this would predict the occurrence of creole-like errors during the acquisition of noncreole languages.

And indeed we find similarities between creoles and acquisition data in all three of the areas surveyed. Typically the same verb can be used both causatively and noncausatively: *we plant the tree* versus *the tree plant;* sometimes even nouns are turned into causative verbs, as in *we cobweb* ("sweep, remove cobwebs from") *the house* (Guyanese Creole) and *we lawnmower the yard* (Hawaiian Creole English). Negative sentences in

English-lexified creoles invariably take the form Subject+*no*+verb phrase. Questions retain affirmative order and are marked only by intonation, plus occasionally an interrogative particle. In other words, it is precisely where creoles differ from English that we find the most persistent problems for children acquiring English (an isolating language that should be almost as easy to acquire as a creole).

The innate grammar would not distinguish causative from noncausative verbs, since all its processes are across the board, and if a process applies to one verb, noun, phrase, or clause, it applies potentially to all members of that class. Similarly the innate grammar has nothing whatever to say about distinctions between the different types of sentence (affirmative, negative, interrogative); a single order is assumed. It follows that "errors" will cluster at points where the target language has introduced modifications of or additions to the innate grammar.

The claim that children are adapted to learn a language of a particular kind has been highly controversial since its introduction. It flatly contradicts the conventional wisdom of both empiricists and nativists— that children are not specially adapted for any kind of language, or that they are equally well adapted for all. It is therefore surprising that virtually no one has tried to test one of the claim's clearest predictions: that creole languages should be acquired more quickly and with fewer errors than noncreole languages. The only studies of which I am aware are by Dany Adone involving the acquisition of Morisyen (Mauritian Creole),: a brief study of the acquisition of WH-movement (Adone and Vainikka 1999) and two books (Adone 1994, 2012), the second of which expands on the first to include experimental work on Seselwa (Seychelles Creole).

Adone's results are highly suggestive. She found, for example, that the TMA system, which follows the general pattern of creole TMA systems, was acquired almost without error. Significantly errors occurred only where Morisyen had added additional TMA markers to the basic three—that is, anterior (past with a current-topic rather than a moment-of-speech reference point), irrealis (futures, conditionals, subjunctives, etc.), and nonpunctual (still ongoing or habitual actions; for additional information on creole TMA systems, see Chapter 8). For instance, *(fi)n*, a perfective, was sometimes confused with anterior *ti*, and the division of irrealis functions between definite-future *pu* and indefinite *ava* led to some further confusion. Children halfway through their third year were regularly producing grammatically correct biclausal sentences:

5. kone kot li 'n ale.
   know where he PERF go.
   "(I) know where he has gone." (Rod, age 2.4)
6. kuma sa pu buze mo pu tonbe.
   when this IRR move I IRR fall.
   "As soon as this moves, I'll fall down." (Ter, 2.4)
7. get lakaz pu rant dan garaz la.
   look house IRR enter in garage the.
   "Look at the (toy) house that can fit into the garage." (Ben, 2.7)

Moreover they seemed almost to be recreolizing creoles: in modern Seselwa, serial-verb constructions have tended to become stereotyped and limited with respect to the number of verbs that can be employed in them, but Seselwa-speaking children used serials much more freely and productively (Adone 2012: 144).

The field badly needs more studies of the present-day acquisition of creole languages, especially of those creoles that have been least influenced by contact with other languages. (Saramaccan is the prime example.) Such studies should be longitudinal as well as cross-sectional, so that the development of particular individuals can be followed over a period of years. The present theory clearly and definitely predicts a series of particular outcomes. Children acquiring a creole language should make fewer errors than children acquiring a noncreole language. Children acquiring a noncreole language make "mistakes" that resemble structures typical of creole languages, but children acquiring creole languages should seldom if ever make "mistakes" that resemble structures not found in creole languages but found in noncreole languages. Such errors that do arise should affect those areas where the creole in question deviates most markedly from the innate instructions for language building. Children acquiring a creole language should reach a final state, corresponding to adult competence, at (on average) a substantially earlier age than children acquiring a noncreole language.

If such studies found any one of these four predictions empirically falsified, this would count against the present theory. If such studies found several or all of these predictions falsified, the theory would find itself in deep trouble.

Conversely, of course, if the predictions are borne out, it would provide extremely strong support for the theory. One strong source for optimism is the so-called syntactic spurt (Radford 1990). Starting at or shortly after the beginning of the third year, children (at least those learning English) typically develop a wide range of structures, including biclausals. In many children this process is extremely rapid. Seth went from *Putting boots. Putting boots on. Put boots. Daddy boots on* to *I don't know where Pink Eddie go, Didja go the beach and see the birds, Didja give the bread the birds* (these are probably not questions, as noted re example (2)), *Daddy made it go ding, I forget to push a light, I like stay in the cemetery* in a period of less than three months. Experimental work by Crain (1991) and others has shown that by age three, children can correctly produce a much wider range of structures than they have ever had occasion to use in spontaneous speech.

A number of explanations have been advanced for this phenomenon, usually involving some delay in the maturation of UG (Radford 1990; Borer and Wexler 1987, 1992), which prevents the full range of structural options from being processed by children. However, both empirical data (Slobin 1985) and theoretical issues (Boser et al. 1992; Deprez and Pierce 1993) support the "continuity hypothesis" that UG is fully present throughout development. From the present viewpoint too there seems no need to postulate any form of grammatical maturation. The syntactic spurt begins immediately after the start of the "lexical" or "vocabulary spurt" (Goldfield and Resnick 1996; Anisfeld et al. 1998), an increase in the pace of word learning. It looks as though the only thing preventing the child from producing longer and more complex structures prior to this vocabulary spurt is simply the absence of the numbers and kinds of words that would be required to fill all the slots in those structures.

Data from "normal" circumstances—in this case, acquisition of well-established languages from abundant, regular, and homogeneous caretaker input—is seldom adequate in itself to resolve scientific questions. Unfortunately experimental methods, which in this case would involve raising children with varying degrees of input deprivation, are ruled out on ethical grounds. We have to turn to "natural experiments"—abnormal circumstances under which input to children has been distorted, degraded, or limited in any way. There appear to be only two such circumstances: those that involve "feral children" (children who for one reason or another have had little or no contact with other humans) and those in which creole languages are created. The former are rare, poorly and often misleadingly documented, and involve considerations (such as a possible "critical period" that makes primary acquisition impossible after a certain age) that are irrelevant to any determination of the way language is structured. Accordingly such cases will not be dealt with here.

CHAPTER 8

Creolization

Creole languages have long defied precise definition. They are generally understood to be relatively new languages, formed mostly (but by no means necessarily or exclusively) through contact between speakers of European and non-European languages. While most such languages take most if not almost all of their vocabulary from a single European language (their "main lexifier"), their grammars contain numerous un-European features. These languages do not fit easily into traditional categories. Consequently almost anything else that can be said about them is controversial to some extent.

According to one view (Hall 1966), creoles originate as part of a "pidgin-creole cycle": typically Europeans come into contact with non-Europeans, and a crude contact language (a pidgin) develops. A pidgin is no one's native language. However, if it is acquired by children, it quickly becomes regular and develops all the resources of a full human language. The fate of the resultant creole language varies. If it remains in contact with its main lexifier, it may "decreolize"—reduce grammatical differences between it and its lexifier—until it is hardly distinguishable from a nonstandard dialect of the latter. If it loses contact, it will be subject only to the processes of linguistic change that affect all language. Although the idea of this cycle has been criticized recently (Mufwene 2001; DeGraff 2003, 2005; Ansaldo and Matthews 2007), the only creole whose full history and prehistory is known in detail, Hawaii Creole English (HCE), followed the cycle through all its stages, and no sufficient reasons have been shown for supposing that other plantation creoles developed differently.

Since Coelho (1880–86) it has been known that creoles resemble one another much more than one would expect for languages developing in different parts of the world, with speakers from a variety of genetically and typologically different languages. Focusing on these similarities rather than differences has been criticized by Mufwene (2001: 51–52), since "there is no compelling reason for downplaying differences in favor of similarities. . . . Both are equally significant." This can be true only if one's interests are purely taxonomic. Anyone with scientific (i.e., explanatory) goals will concentrate, here as elsewhere, on what are the most unexpected aspects of any phenomenon, hence those most in need of explanation. In this case, given the (at least) hundreds of languages of widely varying typologies spoken by the parents of the first creole speakers and the lack of contact between the regions involved, similarities between creole languages are unquestionably the most unexpected phenomena.

Coelho (1880–86) had suggested that similarities might result from universal psychological tendencies, mysterious at that date. After Chomsky it seemed natural to find their source in some form of innate grammar (Bickerton 1974, 1984b). It was for this reason that I first proposed the notion of a biological program for language, the Language Bioprogram (henceforth LB; Bickerton 1981: 133). Thus viewed, creoles were a series of natural experiments with varying degrees of input deformation and deprivation, the results of which might tell us what is innate in language and what has to be learned. As with all natural experiments, those results are noisy and subject to multiple confounds. But since the only way to obtain clearer results would entail limiting linguistic input to different groups of children—a type of experiment ruled out by ethical considerations—those results are the best we are likely to get.

Although the theory presented here differs in a number of ways from the original Bioprogram (it is now firmly rooted in evolutionary theory, among other things), the predictions it makes about creole languages do not differ substantively from the original predictions. Since children start to develop syntax with little regard for the language they are supposedly "learning," and since their (pidgin) input contains little that would add to or run counter to their innate algorithms, they produce similar structures worldwide despite wide variety in the developed languages spoken around them. Since nature has equipped them with mechanisms capable of generating a complete language, the process of creolization can be completed in a single generation. Since once a language has been learned, it is no longer possible to directly access those mechanisms, only children can produce a creole language.

Once one starts looking at creoles from a universal perspective, the problem of defining them changes. Languages described as creoles do not form a natural class. Plantation creoles differ structurally in some but not all ways from fort creoles (languages that emerged from in situ trading contacts between Europeans and non-Europeans in Africa and Asia), which in turn differ from late-creolized pidgins (typically Pacific pidgins that acquired native speakers only after several generations of use by adults). These differences have nothing to do with peculiarities of the areas in which they developed and everything to do with the very different nature and extent of contacts between the participants involved. It is therefore a waste of time to look for "a theory of creolization," if by that we mean a theory that will provide a single explanation for all languages that have been described as creoles.

In contrast, plantation creoles do form a natural class, creating as good a natural laboratory as one could reasonably expect. The uniformity and ubiquity of the economic processes involved in plantation-based societies, described below, remove many variables from the context. Those variables that remain—ethnicity of the population, demographic profiles, length of the various phases of the plantation economy—are instructive rather than confounding, once one grasps the notion of a continuum of creoles.

The Continuum of Creoles

A continuum of creoles must be distinguished from a creole continuum, although the processes that underlie both are the same. In a creole continuum (DeCamp 1971; Bickerton 1971, 1973) such as is found in Jamaica, Guyana, and Hawaii, one finds instead of a single relatively homogeneous grammar a (more or less) continuous spectrum of grammars, ranging from one extreme, largely unintelligible to naïve English speakers to another, barely distinguishable from dialectal forms of English. (See Bickerton 1975/2009b for a detailed description of one such spectrum.) A creole continuum arises because in some creoles, at different periods or different social levels, speakers have had different degrees of access to the lexifier, and the differing varieties that resulted have persisted into the present. Under such circumstances, the gap between what is provided by input and what is required by a full human language—hence what has to be drawn from UG and from preferred strategies for areas underspecified by UG—will inevitably vary, requiring more repair in some varieties, less in others.

If this is true even within single creoles, it must be true a fortiori of creoles as a whole. No two creoles have the same historical profile. They vary in dates of first contact, rates of demographic change, percentage of lexifier speakers, and other factors, so that main-lexifier accessibility and the consequent width of the input/full-language gap will vary from one creole to another. While there is no objective metric that can measure the structural distance between any two languages, crude estimates based on the number of mismatches between creole and lexifier structures suffice to indicate roughly where any given creole should be placed on a continuum that extends from creoles differing radically from their lexifiers to ones differing comparatively little, while others occupy intermediate positions. Clearly the first of these groupings—creoles that differ radically from their lexifiers—holds the greatest linguistic interest, since these should come closest to the output of a pure UG and show the largest component of unlearned strategies.

The notion of a continuum of creoles affords a potential means for falsifying the present theory, which entails an implicational relationship among creoles. Provided we restrict ourselves to plantation creoles, it should be possible to arrange them in a cline such that, assuming the leftmost member has the largest set of nonlexifier grammatical features, the one to its immediate right has a subset of those features, the next a subset of the second's features, and so on until the rightmost member, which would have the smallest subset. If it proved impossible to arrange creoles in this way, that would count against the theory. Moreover the cline should match the sociohistorical data: creoles from areas that have had more lexifier exposure should always be found to the right of creoles that have had less. Any reversal of this relationship would also count against the theory.

The remainder of this chapter will discuss only plantation creoles, and the term *creole* will be taken as referring to these exclusively, unless otherwise stated. Plantation creoles (taken as including languages spoken by those who escaped from plantations) will include all of the most radical creoles, and the circumstances that created them reduce (while of course they cannot eliminate) sociohistorical variables to levels below those of any other type of creole. Similarly *pidgin* will refer to the immediate precreole type of protolanguage, as empirically demonstrated in Hawaii and reconstructable for other creoles—what is sometimes referred to as "jargon"—rather than to stabilized and grammaticized varieties.

The Plantation Cycle

Plantation creoles resulted from a socioeconomic cycle that played out, with minor variations, on a number of islands in the Caribbean, the Gulf of Guinea, the Indian Ocean, and (in one case) the Pacific, as well as in a few isolated parts of South America, all of which were originally uninhabited or (except for Hawaii) underinhabited. Original populations were small and included roughly equal proportions of European (mostly monolingual) and non-European (usually multilingual) speakers (Baker and Corne 1982). Under these circumstances, non-Europeans had a good chance at second-language acquisition of the dominant European language.

The picture changed when the higher profitability of sugar caused a shift from small homesteads to large plantations. A labor-intensive industry and the need to quickly recover initial investment costs caused massive and rapid influxes of slave labor. During this phase, importation of slaves into Suriname over a twenty-year period increased seventeen-fold (Arends 1995), and Jamaica experienced a ninetyfold increase in its non-European population in less than thirty years (Kouwenberg 2009), while in Haiti the non-European population, which increased twentyfold in its first forty-year period, grew two-hundredfold in little more than a century (Singler 2006). Over similar periods the European population seldom more than doubled or trebled.

The linguistic consequences of this rapid expansion were massive and have been misinterpreted by many creolists who assume that if a contact variety of the main lexifier was created in a colony's earlier years, that variety would have been handed on to the newcomers. Some progressive "basilectalization" (Chaudenson 1992, 2001; Mufwene 1996a, 1996b) might then have moved the creole further from the lexifier, due to more imperfect learning by each wave of immigrants. No empirical evidence for such a process has been produced. In reality the commencement of the expansion stage was so sudden and its pace so rapid that gradual degradation of the initial contact language is highly unlikely. The latter would have been as impenetrable to new arrivals as the lexifier itself. But those entrusted with "seasoning" the newcomers and explaining the work program could not afford to wait for them to learn it. To communicate with new arrivals (or for those new arrivals to communicate with one another), all parties had to revert to a rudimentary pidgin, the medium that would serve as the main input for the creole that would eventually develop. From Hawaii there is abundant empirical evidence (Roberts 1995, n.d.) that rapid expansion led to the complete destabilization of a previous semistabilized and expanding pidgin (based not on English but on vernacular Hawaiian) that was much more widespread than Pidgin English prior to 1876. In all other creole societies there is little or no direct evidence of the precreole phase, enabling "pidgin denial" to flourish (see below).

Thus a rudimentary pidgin formed a significant part of the input to locally born children. Another part consisted of the (several, at least) non-European languages that would continue to be used alongside the pidgin by the foreign-born. But the notion that children born in-colony acquired sufficient control over parental languages to impose structures from those languages on the developing creole, although widespread among creolists, has little support from the sociohistorical data.

Consider the life cycle of the average slave child, bearing in mind that "there is now a scholarly consensus that enslaved women played a key role in the field labor force of British Caribbean sugar plantations" (Brereton 2005: 144). Consequently women were brought back into the field as quickly as possible after childbirth, leaving other women to breastfeed and look after their children—women to whom those children often became more attached than they were to their mother (Hindman 2009: 424). In plantation nurseries with children from different ethnic groups, pidgin would have been the only means by which all the children could talk to one another throughout the most critical phase of their linguistic development. Subsequently from age five onward, children worked in the "third gang," employed on light cleanup work (Higman 1995: 162). Here their main interaction would have continued to be with other children. They would see their parents (if either or both survived) mostly on

Sundays and holidays. It follows that children raised under plantation conditions would have needed the pidgin more than they needed their parents' language(s).

The expansion phase would not end until most if not all of the cultivable land in the colony had been brought into production. At that point populations would stabilize, with owners importing only enough slaves to make up for negative population growth. By this stage a developed creole language would already be in place, and subsequent immigrants would learn it as a second language.

In light of the foregoing, it is strange to be informed that "for many other creolists, the scenario for creole genesis that the Language Bioprogram Hypothesis requires is crucially at odds with established facts about the history of the colonies where the creoles arose" (Kouwenberg and Singler 2008: 4). The authors fail to cite a single alternative scenario from these "other creolists" or even one of the "established facts" on which their opinions are allegedly based. Since a more detailed account of two of the most crucial sites for creole creation, Hawaii and Suriname, will be given later in this chapter, readers can judge for themselves which scenario comes closest to the facts. The claims made in Bickerton (1981, 1984b)—that creoles are produced by children, in a single generation, from a relatively structureless early-stage pidgin, with the aid of an innate program for language—have been challenged frequently but never convincingly and often in defiance of empirical evidence.

Criticisms of the Innate Program

Before looking at specific criticisms, it is necessary to dispose of certain confusions for which I myself am largely responsible. The severity of these is revealed by Baptiste (2012) and McWhorter (2013). These involve the relationship between the LB and Chomsky's Universal Grammar.

Clearly Chomskyan UG was the stimulus for the LB without which the LB could not even have been hypothesized. In the beginning (Bickerton 1974) I was less interested in what the LB was than in what, on the creole evidence (those areas in which plantation creole grammars showed a higher degree of similarity than existing theories predicted), the LB contained. By 1981 I had formed a vague notion that the LB represented a single monolithic grammar, but how this related to the Chomskyan model remained mysterious. In commentary on Bickerton (1984b), several generativists pointed this out. Recall that the paper in question was published at a time when the principles-and-parameters model of UG (Chomsky 1981), a stunning achievement in conceptual terms, had just appeared and was enjoying enormous prestige. Obviously there could not be two Universal Grammars. But there appeared to be a simple way the two models could be reconciled. Creoles might represent the set of unmarked parameters in a parameterized UG, and if "unmarked parameter" meant "default parameter," that setting would be assumed by the child in the absence of evidence to the contrary (Bickerton 1984a, 1986). Since in most cases contrary evidence from a pidgin input would be too limited and conflicting to establish settings, the set of default parameters would apply repeatedly in different creoles.

At this point (in the late 1980s) the major focus of my interest shifted to the evolution of language, and from then on anything I published on creoles was purely reactive. In fact, as I realized only recently, this was a serendipitous development. It turned out that one could not hope to fully understand how creoles emerged without understanding how language had evolved, and understanding how language evolved suggested that the true nature of syntax (a mix of biologically given and inductively learned materials, as described here) differed from any account hitherto proposed. Among the benefits of this new account was a final clarification of the true relationship between the LB and UG.

The two cannot coexist: they are different theories about the same thing. The LB, as now conceived, is the single monolithic grammar originally suggested but with a much narrower scope, as Chapter 5 showed. As such, it stands in stark contrast to rich versions of UG such as the principles-and-parameters model, but much less so to more impoverished models within the Minimalist Program, such as those of Hornstein (2009) and Boeckx (2010). The LB is used by children in creolizing situations in exactly the same way as it is by children in "normal," established-language settings (see Chapter 7). All that differs is the quantity or quality of primary linguistic data in the two cases. Consequently creoles consist of:

1. The universal algorithms for phrase and clause construction.
2. Minimal repair strategies for areas left unspecified by those algorithms.
3. Any substrate or superstrate phenomena sufficiently salient and consistent in the pidgin data.

It is no longer assumed, as it was in Bickerton (1984a), that the LB might represent unmarked settings of parameters; the notions of "parameter" and "markedness" are themselves no longer relevant.

Variation in the amount of contribution from (3) is, of course, what causes both the continuum of creoles and the continuums within some creoles. That amount is easy to overestimate. Take, for example, serial verb constructions in creoles, found in a wide range of substrate languages and widely regarded as due to substrate influence. In fact such constructions follow naturally from the innate algorithms. Since these do not specify any limitation on the content of head or modifier, verbs can freely modify other verbs. If possession of a serializing substrate was a necessary and sufficient condition for creole serial verb constructions, such structures would not be found in the Indian Ocean creoles (where their presence is nowadays taken for granted by native-speaking linguists; see Adone 2012; Syea 2013), since these have no serializing substrate languages.

Where repair strategies are undertaken, possibilities may be starkly limited in number—in extreme cases, to just one. It is hard to see how a set of question words could be re-created, as they are in Sranan, Haitian, and other plantation creoles, save by taking any interrogative form that could be salvaged from the pidgin and attaching it to words indicative of time, place, person, thing, and so on. In other cases, a single strategy may outvalue any alternative. For instance, in the TMA system (see below) only one logically possible system combines a minimum of lexical material with a maximum of explicit semantic coverage. The source given in this chapter seems more plausible in evolutionary terms than the explicitly language-dedicated mechanisms suggested in earlier treatments.

It cannot be too strongly emphasized that none of these changed assumptions about the content of the innate component makes the slightest difference to the original claims of the LB as listed earlier. Each of these claims has been repeatedly challenged over the past four decades. It has been maintained that adults, not children, were the main agents of creolization. Creole creation is said to have taken more than one, possibly several generations. The very existence of precreole pidgins has been denied. The differences between creoles are supposedly so great that no innate mechanism, even if one existed, could have played a significant role in their emergence. Where features in creoles differ from those in their European lexifiers, differences are attributed to the continuing influences of substrate languages (the languages spoken natively by the ancestors of creole speakers).

In evaluating these challenges, it is important to recognize that underlying them is a mind-set common among creolists and far from unknown elsewhere. According to this mind-set everything has to come from somewhere else. Every structure and feature in a creole has to have a source in some already existing language that is somehow involved in the creolization process. Consequently those sources have to be found, and once they have all been identified, creolization has been explained. Perhaps possessors of this mind-set have not realized that, if they are right, nothing new can arise or can ever have arisen in language. This would give us the choice of two equally bizarre conclusions: that language has always existed in its present form, or that it doesn't exist at all. The following sections deal in turn with five main claims made by critics of the LB.

Adults versus Children

Since Alleyne (1971) raised the issue, it has seemed to many implausible that the vast majority in a plantation colony should have soldiered on with an unstructured pidgin for the several decades it undoubtedly took for the locally born to equal them in numbers. But now that Roberts (n.d.) has amassed a database of over two thousand citations of the various contact languages spoken in Hawaii since the 1780s, we have abundant empirical evidence that what seemed bizarre to Alleyne and others is simply what happens when adults without a common language have to communicate with one another.

Pidgin English spoken in Hawaii in the decade from 1910 to 1920, three or four decades after the start of mass immigration, when the first creole speakers had reached or were reaching maturity, was no more regular, structured, or expanded than Pidgin English spoken in the closing years of the eighteenth century. Typical examples include the following (all verbatim evidence from court records):

1. What matter you speak buy wine Japanese store, then me *lele pelekia* (1911).
2. Me buy, me go house *kaukau kela opiuma* shit (1911).
3. You license no all same name? (1913).
4. Long time shoe no finish (1914).
5. He son-a-bitch, no good. I fix him, by and by fix (1915).
6. Chong Chin strong man he *hanapaa* over there, neck *liilii koe make* no (1917).
7. Yes, no can help, no can hide (1918).

After 1920 more and more new immigrants acquired the emerging creole as a second language with varying degrees of success (this was long before locally born speakers reached numerical parity with immigrants). Pre-1920 arrivals continued to use the variable and virtually syntaxless pidgin for the rest of their lives. (Examples from the early 1970s are given in Bickerton and Odo 1976; Bickerton 1981.)

It is true that in the Western Pacific, Tokpisin, Beach-la-mar, and other pidgins developed systematic structures closely approximating those of a full natural language with (apparently) little or no contribution from children. However, in these cases the processes took several generations. Moreover there existed conditions none of which applied in Hawaii or in other plantation colonies: speakers had similar substrate languages and they continued to live in their ancestral homelands where those languages flourished. Consequently we can assume that most Western Pacific speakers were fluent bilingual adults with their ancestral language dominant, which would have made possible massive transfers from substrate languages. We cannot assume the same for Hawaii. As will be shown, there are indications that children's competence in substrate languages was too poor for them to control anything more than the simplest structures, if those.

The only group to show consistent and significant substrate influence in Hawaii was the only group raised purposefully as Japanese-dominant bilinguals. This group, known as the *kibei*, consisted of Japanese who were born in Hawaii but who, usually between ages five and nine, were sent back to Japan to complete their education. On returning to Hawaii, *kibei* spoke a variety of HCE that showed strong Japanese interference in such gross details as SOV word order and head-final relative and adverbial clauses (Bickerton 1977), features found elsewhere only sporadically in the speech of older and more traditional Japanese. The clincher is, of course, the empirical finding by Roberts (1998) that in Hawaii typically creole forms and structures came exclusively from locally born individuals, most of whom were explicitly described as "children" or "adolescents" by those who cited them.

Sole justification for the mantra adopted by DeGraff (1999) and a number of subsequent writers—"Adults innovate, children regularize"—lies in the fact that some typically creole words (e.g., TMA markers *bin* and *go* or purposive marker *fu* from *for*) were used, if only occasionally and unpredictably, by pidgin speakers. Use of the pidgin lexicon by creole speakers is inevitable; they have to get words from somewhere. But pidgin speakers never used *bin* as an anterior rather than a simple past marker or *go* exclusively as an irrealis rather than a mere sequence-of-events marker. Moreover, at the level of structure as opposed to individual words, there is no "adult innovation" at all. Three structures nowhere found among immigrant speakers are relative clauses ((8), finite purposive clauses with *for/fu* (9), and resumptive pronouns obligatory with indefinite-reference subjects; (8)–(10) are all attested examples). However, all three are structures frequently produced by creole speakers.

   8. The guy gon' lay the vinyl been quote me price.
      "The guy who was going to lay the vinyl had quoted me a price."
   9. More better for I write the answer.
      "It's better for me to write the answer."
   10. Plenty old Hawaiians they do that.
       "A lot of old Hawaiians used to do that."

In other words, there was nothing for children to "regularize." They innovated and regularized at the same time, unaided by adults.

Support for an "adults first" model of creole creation has been drawn from the supposed fact that proportions of children in plantation colonies (Haiti was the most frequently referenced) were far too low to create a viable language for the entire community (among many others, see Goodman 1985; Singler 1986). Based on a study of the original French census records in Aix-en-Provence, Bickerton (1990a) showed that at no time in Haiti's early history did the number of children under twelve fall below 20 percent of the total population. (Indeed the lowest figure found in any plantation colony was 13.8 percent, in early Suriname.) For comparison, the CIA's *World Factbook* gives the current proportion of children up to age fourteen in the U.S. population as 20.2 percent. Despite these facts, Lefebvre (1993: 259) continued to insist that "there was an almost total absence of reproduction among the population during the critical period."

These facts support something that is obvious to any lay person who has had occasion to compare the linguistic abilities of adults and children. They will have noted, for example, that if a parent and child move to an area where a foreign language is spoken, the child will, in the vast majority of cases, acquire that language much more quickly and fully than the parent. Data from Hawaii show that this difference in learning ability is far exceeded by the differences in creative ability. In any plantation society there would have been locally born individuals speaking a fully developed human language while most foreign-born adults were still struggling to stabilize a workable means of communication. Inevitably under these circumstances, foreign-born would seek to imitate locally born and would pick up some second-language version, adequate or otherwise, of the new creole. That is what happened in Hawaii by the time the first creole-speaking generation had become adults (even though they still represented only a small minority of the total population). In the absence of indications to the contrary, we must assume that similar developments took place in other plantation communities too.

Despite the fact that all empirical evidence indicates children rather than adults as the creators of creoles and only theoretical conjectures support the claim that adults are responsible, the belief that creoles are created mainly by adults has broadened and strengthened in recent years (among others, see, e.g., Mather 2006; Plag 2008a, 2008b; Siegel 2009). The main support for this belief is based on similarities between structures found in substrate languages that resemble structures found in creoles.

The Influence of Substrate Languages

The existence of any kind of UG constrains the overall form languages can take. It is perhaps less obvious, if equally true, that even a skeletal UG such as that described in these pages constrains the form of languages to a surprising extent. Granted, what such a UG leaves unspecified results in wide variation among languages. But much of that variation is localized in particular areas of the grammar, and the number of ways to fill the gaps is quite limited in number. This means that some close syntactic matches are highly likely to be found even between any randomly chosen pair of typologically similar (e.g., isolating) languages.

At first glance the number of matches between creoles and their substrates might seem too high for coincidence to provide a valid explanation. However, it is also the case that, apart from the Indian Ocean creoles (Morisyen, Seselwa) and HCE, all creoles have had a mostly West African substrate. In the creole literature it is generally assumed that West African languages have a relatively homogeneous structure. Unfortunately Africanist linguists fail to share this perception.

Consider what are often treated as the jewels in the crown of substratist argument: serial verb constructions (SVCs; e.g., Sebba 1987; McWhorter 1992; Migge 1998). Typical of the substratist approach is McWhorter (1997: 154–155), who states, "Serial verbs are typical of Caribbean creoles because of their predominance in *West African languages.... All African speakers* in a given context would have been accustomed to generating and processing language through them" (emphasis added).

However, a specialist in African language (Dimmendaal 2001: 382) complains, "A stereotypical view of African languages sometimes encountered in the general literature is the presence of serial verb constructions. In actual fact, this phenomenon has a rather restricted distribution both genetically and areally. It is found in a largely contiguous zone stretching from the Ivory Coast to Nigeria, in languages belonging to different subgroups of Niger-Congo: Kwa, Western Benue-Congo and Ijoid." In light of this statement, consider the fact that "European slaving forts or slave-trading depots extended from Cape Verde in the Senegambia south to Benguela in the Angola region" (Richardson 1992: 65). This means that the catchment areas for slaves included those inhabited by speakers of Bambara, Wolof, Hausa, Fulani, Kikongo, Lingala, and many other Niger-Congo and Bantu languages that do not serialize.

We do not have full demographic profiles of slave provenance over time in each Caribbean colony, so it is impossible to determine the proportion of speakers of SVC languages at any given time or place. However, it is highly unlikely that they predominated at all times and places and highly likely that in many areas speakers of languages without SVCs predominated. If that is so, it becomes mysterious how SVC speakers managed to get non-SVC speakers to adopt SVCs.

Mysteries are dissipated when it can be shown they are based on false facts. All three of the creoles that lack a West African substrate possess or have possessed SVCs. Morisyen and Seselwa, universally held to be SVC-free (Bollee 1977; Corne 1977; Papen 1978; Muysken 1988) was shown to have a wide range of SVCs (Bickerton 1989, 1996) that previous observers had inexplicably missed. HCE had directional serials (V *come*, V *go*) in its earlier stages (Roberts 2004, citing contemporary sources). The fact that it doesn't have any nowadays is unsurprising in view of the fact that at least one other creole that does have a West African substrate (Bajan, spoken in Barbados) doesn't have SVCs either. In both Hawaii and Barbados the most plausible reason for this is the extent of pressure from English on both islands rather than any substrate effect.

This has not stopped some scholars from attributing a variety of HCE features to substrate influence. Siegel (2000, 2007), following suggestions in earlier work by Roberts, proposed that HCE use of *get* in both possessive and existential senses stemmed from the existence of similar verbs in the substrate (Cantonese *yauh*, Hakka *ju*, Portuguese *ter*, Hawaiian *he* or *loa'a*). Roberts (2004: 254) further proposes that the HCE expression *no mo(re)*, a negative possessive/existential—

11. Us nomo money.
"We don't have any money."

—derives from a Cantonese verb, *mouh*, of similar meaning. Siegel (2000) points to a locative copula *hai* in Cantonese that parallels HCE *stay*, a verb that both Reinecke and Tokimasa (1934) and Knowlton (1967) saw as originating in Portuguese *estar*. Even the use of *stay* as an imperfective-aspect marker is seen as deriving from Portuguese *estar*+verb, as well as Cantonese *hai/haidouh*, not to mention Hawaiian *noho* (Roberts 2004:

257–258). Similarly Siegel (2007) claims that the other two TMA markers, *been* and *go,* correspond to Portuguese uses of *ter* and *ir,* respectively. The purposive complementizer *for* is said to come from Hawaiian *no/na* and/or Cantonese *waihjo* and/or Portuguese *para* (Roberts 2004: 260–261), and even a source for its use with transitive clauses and nominative subjects (as in (9)) has been located by Siegel (2007) in a Portuguese subjunctive construction. Directional serials of the type Verb+*come* or *go,* earlier frequent among children though rare or nonexistent nowadays, are attributed by Roberts (2004: 262) to a Cantonese serial construction, and she even goes so far as to hint at an origin for *us,* HCE's first-person-plural nominative, in Northamptonshire dialect (268).

What both Siegel and Roberts fail to mention is that all of these features occur in many creoles worldwide. Possessive existential *get* (or its French or Portuguese equivalent) is found in Guyanese, Haitian, Papiamentu, and Sao Tomense (Bickerton 1981: 66–67). A negative possessive or existential, *napa*—an exact parallel of *nomo*—is found in the Indian Ocean creoles (Corne 1988, 1995). The morpheme *de* (nowadays contracted to *e* in its aspectual sense) is used, like *stay,* both for [-completive] aspect and locative copula in Sranan. A purposive complementizer derived from *for* is found in virtually all English creoles and its equivalents (with reduced versions of *pour* or *para,* respectively) in most French and Portuguese creoles (Bickerton 1984b); the purposive clauses it heads are usually, as in HCE, finite, contrasting with equivalent infinitive structures in all three lexifiers. Directional serials have been reported from numerous Caribbean creoles. In all these cases, African substrate origins have been repeatedly invoked, but of course the corresponding HCE cases are never mentioned, just as the African cases are never cited when HCE is under discussion. It is clearly time for Atlantic and Pacific substratists to engage one another in serious conversation.

If the forms noted by Roberts and Siegel really came from substrate languages, one would expect them to be present in the production of adults who had full competence in those languages. The fact that not one of these forms occurs in any of Roberts's database citations produced by nonlocally-born adults is therefore truly remarkable. Perhaps the most striking case is that of the supposed derivation of *stay* and its functions from Portuguese *estar.* Siegel (2007) claims "general agreement" on the correctness of this derivation, citing Reinecke and Tokimasa (1934),

Smith (1939), and Carr (1972). He fails to mention, as do they, that *stay* did not enter the Hawaii vocabulary until after 1900, before which the Hawaii Pidgin English locative was *stop*. (Some of my oldest informants were still using *stop* in the 1970s.) Since Portuguese had been present in large numbers for nearly a quarter of a century by 1900, it is surely remarkable that not one native Portuguese speaker in Hawaii (nor anyone who might have picked it up from them) has been recorded as using *stay* in all that time. Moreover *stay* as an aspect marker is nowhere cited before 1920 and is noted in Roberts (1998) as clearly an innovation by the locally born.

This leaves the introduction of *stay* by bilingual Portuguese/HCE speakers (who would have to have been locally born children) as the only route by which *estar* could have influenced *stay*. Later we will see how unlikely that was. Failure on the part of substratists to confront inconvenient facts such as these suggests that substrate theory cannot provide a coherent theory of creolization.

"Gradual" Creolization

The notion that creoles evolve gradually, over several generations, has widespread, if far from complete, support (Arends 1989, 1995; Plag 1995; Mather 2006; but see Smith 2006). However, evaluating the evidence requires much more care than most advocates of gradualness have shown.

First, there is confusion of first-citation dates with origin dates. Taking a historical approach, Baker (1995) uses first-citation dates to show that pidgins and creoles are indistinguishable and that development of creoles often spread over a century or more. He ignores the fact that there is certain to be a time lag between first use and first citation, and that this time lag may be both unavoidable and quite long. For instance, no Sranan text exists between 1718 and 1767, so a Sranan feature originating around the time of the first text but too late for inclusion in it would have had to wait at least half a century before surfacing. He ignores the unhappy concurrence of three facts about early texts and their content: that virtually all were written by nonnative speakers, that many texts are quite short, and that all linguistic features in any language vary in frequency, meaning that some are much less likely to be represented, especially in short texts, than others. (Among these are combinations of two or three TMA markers, invariably found later than single markers in creole texts.) Given these problems, first-citation dates simply cannot provide an accurate record of the pace of historical development.

Second, there is confusion between changes entailed by creolization and changes that involve normal intergenerational language change. Creoles are natural languages; thus they are as subject to "normal" change as any others. Most if not all of the evidence for gradualness shows changes that either alter already existing creole features or produce alternative ways of expressing things already expressed. (See Bickerton 1991 for some detailed cases, specifically answering the arguments of Arends 1989.) Here there is simply a double standard for creoles. English has changed considerably since the seventeenth century (when most creoles originated), but no one has yet suggested that what was spoken in seventeenth-century England still had not managed to become English.

What might seem the most direct evidence for gradualness, from Roberts (2000, 2004), will be discussed at length later. Roberts's claim depends not on any linguistic data but solely on the belief that a child cannot have two native languages. Yet practically all scholars in the field of acquisition accept that more than one language can be acquired natively. Indeed the term *native bilingual* (Guthrie 1984; Thorn and Gathercole 1999; Tse 2001) is standard in the second-language acquisition literature, while Meisel (2004: 95) flatly states, "The most important insight gained from studies on child bilingualism over the past 25 years is perhaps that simultaneous acquisition of two or more languages can indeed be qualified as an instance of multiple first language acquisition."

In short, there is no empirical support for a multigenerational or even a bigenerational genesis of creoles. A one-generational genesis is inevitable if UG and acquisition are as they have been described here. It remains unclear what set of assumptions, if any, would entail the alternatives.

Pidgin Denial

The belief that pidgins did not always (or never did) directly precede and give birth to creoles unites a variety of scholars who sometimes hold diametrically opposite positions on other issues. According to Lefebvre (1986: 283), who believes Haitian is substrate-based, "No pre-creole pidgin phase is postulated," and her view is shared by DeGraff (2003), who believes Haitian is a dialect of French. In their attempt to "deconstruct" creoles, Ansaldo and Matthews (2007), without explicitly denying the existence of precreole pidgins, entail such a denial by their rejection of any "break in transmission" between creoles and antecedent languages.

Unfortunately in the one case where we know from massive and detailed empirical data exactly what directly preceded a creole, we have abundant evidence that this was indeed a structureless pidgin. Moreover, contra the belief that pidgins always constitute an offspring of some particular lexifier—Pidgin English, Pidgin French, and the like—it had a highly macaronic vocabulary (Roberts 1998, 2004, n.d.), that is, one with words drawn from a variety of languages. The objection that other creoles show no evidence of such a stage is baseless. The existence of a *baragouin*—a French word of Breton origin meaning "jargon" or "unintelligible gibberish"—that preceded the creoles of the French Antilles is well attested (Wylie 1995). In the next section we meet a number of words in a variety of creoles that are obvious fossils of a precreole pidgin. Many of the earliest attestations of creoles, such as Herlein (1718), a text that looks like what a second-language learner of Sranan might have produced, or the earliest citations of Morisyen (Mauritian Creole) in Chaudenson (1981), point in a similar direction. Further indirect evidence can be found in the extremely diverse (and loaded with homonyms) vocabulary of the Suriname creoles, which yields little-noted but extremely significant data whose implications are discussed below.

This evidence is consistent with what has been found in pidgins that never developed into creole languages (Bickerton 1999). The lexical sources for words in the oldest recorded pidgin, the Lingua Franca of the Mediterranean, varied considerably across time and space (Schuchardt 1980). Russo-Norsk contains not only both Russian and Norwegian words but words from several other languages (Broch and Jahr 1984). Despite its name, Chinook jargon gets less than half its vocabulary from the Chinook language; the rest is divided among French, English, and several other American Indian languages (Lang 2008). Fanagalo, a pidgin of South African miners, contains a mix of English, Afrikaans, and African (mainly Zulu) words (Mesthrie1989). Bakker (1989) describes a pidgin that began life through contacts between French fisherman and Icelanders, yet most of its words have Germanic sources, from Flemish to Danish. Pidgins are exactly what one would expect them to be if, rather than frustrated attempts to learn some single language, most of them are simply attempts to communicate by any possible means across formidable language barriers. As such, they are likely to have emerged in a structureless, macaronic form wherever large numbers of people speaking mutually unintelligible languages suddenly came in contact with one another.

Pidgin denial resembles Holocaust denial and climate-change denial in that deniers seem somehow able to counterbalance a mountain of empirical evidence with a few trivial and irrelevant factoids, such as that there is no overlap between areas where pidgins are spoken today and where they were hypothesized in the past, or that the term *creole* originally applied to people or animals rather than languages, or that the term *pidgin* came from China (Mufwene 2008, where these are listed among reasons for supposing that the pidgin-to-creole cycle is a "myth" that linguists should do away with). Indeed major papers that seek to demolish the whole notion of a pidgin-to-creole cycle do not even mention, let alone discuss, the case of Hawaii (Mufwene 2003, 2008; DeGraff 2005). When Hawaii is referenced, it is only to make statements such as the following:. "In Hawaii, American missionaries first taught English to members of the royal family, who would later become instrumental as interpreters during the colonization and the economic exploitation of the islands" (Mufwene 2009: 5). The vision of Hawaiian princes resolving disputes between Carolina straw bosses and Chinese laborers has much charm, if zero historical support, and suggests there is no limit to the lengths to which pidgin deniers will go.

Creole Differences

The first thing to note under this subhead is the curious belief apparently held by many creolists that if there were a biological program for language, that program would have to be executed universally and without exception or modification, so that if any creole deviated from the program's specifications this would automatically invalidate the whole notion of such a program. In fact "the genetic program metaphor encourages the idea that each decision point in development or behavior, and each environmental input, acts under a genetic directive or a genetic set of rules, like 'in condition X, do A' or 'in condition Y, do B.' But there are no such rules in the genome, only a set of templates for molecules that will become part of the phenotype. The behavior of those molecules depends as much on the nature of the environmentally-supplied materials that compose them as on the genetic template that contributes to their organization" (West-Eberhard 2003: 14–15).

A simple misunderstanding of how biology works lies at the root of many criticisms of the universalist approach. The "genetic template" cannot even begin to operate in the absence of "environmentally-supplied materials"—in other words, words—since the skeletal UG described in Chapter 5 has to interact with lexical material of some kind. In the circumstances of normal generation-to-generation transmission, the environmentally supplied materials are abundant, sufficient to yield the vast, though far from limitless, diversity of human languages. In the circumstances that characterized plantation colonies, those materials were both sparser and more conflicting, reducing the contribution they could make to the child's intake. Children still had the algorithms for phrase and clause construction that were part of their linguistic heritage, but they could not reliably reconstruct any previous language's system for specifying those areas of grammar that this skeletal UG left underspecified.

Here is where the present theory deviates from previous models. The original bioprogram (Bickerton 1981, 1984b) was cautious in its notions of exactly what the innate component consisted of. The implication was that the bioprogram, like its generative congeners, spelled out, if not everything in grammar, at least a very great deal of it. This still left room for some variation in creole features. The "continuum of creoles" concept elaborated earlier would still apply: the differing amounts of substrate and lexifier features accessible in each situation would inevitably cause differences in the resulting language, as children tried to flesh out the skeletal structures that were all that UG gave them.

Children in a creolizing situation would use those scraps primarily to fill the gaps left by an incomplete UG. The creole had to have word order, and word order wasn't specified, so they would take what they could get in the way of examples, and what they could get would never be quite the same in any two places. Creoles show wider differences in noun phrases than in clauses (Kihm 2008) because the UG order of attachment in clauses partially specifies sequence, whereas in phrases there are no such constraints. But recall from Chapter 6 that there are both underspecifications that must be repaired and underspecifications that may be repaired.

Hence there are two sources for variation in creoles. One, stemming from the continuum of creoles, is variation in the richness or otherwise of material from other languages available to the child, leading most frequently to variations in the closeness of a given creole to its main lexifier. The other source, not considered in earlier treatments, consists of the strategies adopted to repair underspecifications of the second type. This will become clearer if we consider some particular cases of repair. Two of these involve question words and TMA systems.

TMA systems and question words are things languages seem to have to have; no language entirely does without them. Yet they differ in their forms and functions widely across languages—in the case of TMA systems, dramatically so, for (outside creoles) very few languages have identical systems. The essence of these things must, in some sense, be innate. It is hard to imagine a language that had no means of indicating whether utterances were statements or questions and, if the latter, what exactly they were questioning. It is hard to imagine a language that lacked means to express the relative timing or reality of events. But exactly how or why these things are innate is harder to say. There seem to be a set of human presuppositions about language that is an emergent property of the kind of general cognition we have. However, there are far more ways of making a TMA system than there are of making question words. If you lose your original question words, there is no obvious way to replace them other than by choosing a single morpheme that will simply mean Q—"This is a question!"—followed by another word that specifies what kind of question it is: a location in "Where?" questions, a general term for person(s) in "Who?" questions, and so on. All creoles that do not preserve question words from their main lexifier replace them with bimorphemic question words in precisely this way.

However, choice of both what to use as the key question word and what to use as its descriptor are not determined by anything except overall semantics. "Who," "what," or "which" may be chosen for the Q item. For the second item, "where" questions may use "side" or "place," "when" questions may use "time" or "hour," and so on. Differences may arise even between closely related creoles. "Why" is *san-ede* (literally "thing-head") in Sranan and *andi-mbei* (literally "what makes") in Saramaccan.

The TMA system is more complex, although we find virtual identity between some totally unrelated creoles, for example, Saramaccan, spoken in Suriname, and Fa d'Ambu, spoken on an island off West Africa (Bakker et al. 1995), as well as a common inner core in all creoles. Assuming that general cognitive constraints create the need to express the time, reality, and completeness of actions and events, this core system is the most parsimonious possible, yielding three markers (one for each semantic category) and, since no stipulations limit combinability, a total of eight "tenses."

The relatively slight differences between creoles found in this system derive from two causes. Children exposed to a pidgin will try to make use of anything they find there. Subsequently they may or may not incorporate such material into their TMA system. In Hawaii Pidgin English there were two sequence markers, *baimbai* and *pau*, the former used for both next-in-sequence and future events, the latter for completed events. (*Pau* means "finish" in vernacular Hawaiian.) There are grounds for supposing that markers with these functions recur repeatedly in pidgins, and in a number of cases are incorporated into their subsequent creoles. The next-in-sequence marker is clearly a candidate for expressing irrealis modality, while the completed-event marker is a natural for completive aspect. In Hawaii neither marker was incorporated into the creole TMA system, though both remain in their original phrase- or clause-final position as sporadic adverbs.

However, other creoles (including some that are not plantation creoles) behave differently. In Tokpisin we find *baimbai* reduced to *bai* in its creolized form and incorporated, and *pinis* (from "finish") incorporated but left after the verb. In Portuguese-related or Portuguese-influenced creoles, we find *kaba* (from Portuguese *acabar*, "finish") as well as *lo(go)* (Portuguese *logo*, "subsequently, soon"). In French-related creoles, we find *fini* (French *fini*, "finished"). In English-related Atlantic creoles we find *don* (from "done"). In Papiamentu, *lo* (reduced from *logo*) was incorporated as an irrealis marker but remained in sentence-initial position. In Suriname creoles, *kaba* was incorporated as a sporadically recurring completive but remained in final position. In French-related creoles, *fini*, frequently reduced, moved into canonical TMA position to compete with anterior *ti*. In English-related creoles, *don*'s fate was variable.

The second cause of deviations has nothing to do with creolization at all but involves subsequent linguistic change. Modern Sranan has two irrealis markers, *sa* (either from Portuguese *saber* or Dutch *zal*) and *(g)o*, but according to Winford and Migge (2007: 78) "all of the tense-aspect categories of the modern [Surinamese] creoles, with the exception of Future *(g)o,* are already well-established in early texts." In other words, Sranan somehow picked up *go* subsequent to creolization, and possession of two irrealis markers led to division of meaning, with *(g)o* and *sa* marking more and less probable events, respectively. Guyanese similarly picked up *doz* (from *does*) as a marker of habituality, yielding a contrast between the variety of Guyanese furthest from English (11) and varieties somewhat closer (12):

12. i a wok haad.
    "He is working/works hard."
13a. i a wok/ i wokin haad.
    "He is working hard."
  b. i doz wok haad.
    "He works hard."

If we leave out of account pidgin fossils and postcreolization changes, the only differential in creole TMA systems is the extent to which, if at all, more extended and/or intensive contact with the lexifier has led to the replacement of the typical creole markers and the semantic categories they represent by main-lexifier forms and categories. In other words, such differences are the inevitable products of unavoidable facts: creoles have to get vocabulary from somewhere, they have to repair some of UG's underspecifications, and no two creoles will have exactly the same input from other languages. In no way could these facts constitute evidence for rejecting the universalist approach to creoles.

To sum up, none of the criticisms leveled at the universalist approach has proven valid. Notwithstanding this fact, it has been claimed repeatedly that the approach has been finally and decisively refuted. One can only wonder about the basis for such claims, since few who denounce the LB even mention, and hardly any seriously discuss, let alone answer, a long series of papers (Bickerton 1984a, 1987, 1988, 1990a, 1991, 1992a,

1994, 1995, 1996, 1999, 2006) that have already answered these and other criticisms.

Perhaps the two most crucial areas in the study of creoles are Suriname and Hawaii. Suriname is crucial because, unlike most areas, it contains not one but several creoles, and these are further from their main lexifier (English) than any others. Hawaii is crucial because it is the only area in the world where we have massive empirical evidence not only for the creole itself but for the stages that preceded it. I show here that the present approach can give a better account of these two areas than any competitor.

Sranan and Saramaccan: Common or Separate Origins?

Suriname was first settled in 1651. Forty years later it had split into two communities. One consisted of plantation owners and slaves and developed a language known as Sranan. The other consisted of maroons (escaped slaves) and developed a language known as Saramaccan. These languages are so similar in their grammars that it has been assumed by virtually all specialists in the area that they are offshoots of a single creole language (or at worst, a stabilized and fully developed pidgin well on its way to becoming a full language). If, however, there was still only a structureless and macaronic pidgin in 1690, then while the two languages must have developed from this same pidgin, their similarity in structure can arise only from the fact that the same universal factors discussed in this chapter operated twice, independently, to produce in each case highly similar if not quite identical results.

Evidence to support the latter scenario comes from two distinct sources. One is the history and demographic profile of Suriname between 1651 and 1690. The other is a comparison between the vocabularies of Sranan and Saramaccan, which has not previously been undertaken. These two quite different sources complement one another, both strongly suggesting that Sranan and Saramaccan had separate births and that their structural similarities form powerful evidence for the universalist position.

The Early History of Suriname

Suriname started around 1650 as primarily an English homesteading colony such as Barbados originally was, with at most a 2:1 ratio of slave to free. Starting in 1667 it underwent a series of traumatic events: a Dutch invasion and ultimate takeover, the loss of most of its entire original population when the English and their slaves left en masse, various plagues, and Amerindian attacks. According to Migge (1998: 221), "Roughly three-fourths (1670s) to around half (1680s) of all slaves . . . died or left the plantation." In the period 1680–1700, the arrival of seventeen thousand slaves (Postma 1990) led to a population increase of less than eight thousand (Arends 1995).

According to Price (1976), a 1690 slave revolt on the Machado plantation (one of a number of Portuguese-owned plantations) initiated a division between Sranan (the language of those who remained on the plantations) and Saramaccan (the first of several languages of those who escaped). But in the absence of surviving texts, the linguistic situation in 1690 remains unknown. The history of the preceding decades lends little support to any scenario of early creolization. From 1667 on, English proprietors left the colony with their slaves, and by 1675 there remained only about fifty of the original slave population. Dutch proprietors were slow to take advantage of their new colony, so that in 1679 there were still only a thousand slaves (down from three thousand fourteen years earlier). What this means is that there was an extremely narrow window, around five or six years at the most, for arriving slaves to fully acquire whatever contact language the departing slaves had developed. Immediately afterward the newly arrived slaves would have had to transmit the same contact language, in all its details, to incoming slaves while (if the figures cited above from Migge 1998 are correct) they were losing between 5 and 8 percent of their own number *every year!* At the same time, this rapidly diminishing band of "window" slaves was being swamped by more and more new immigrants, since, as Arends (1995: 269) observes. "during the first fifty years of colonization, *the entire black population*" (in which the surviving window slaves formed already a shrinking minority) "was outnumbered by arrivals from Africa *every three to five years*" (emphasis added). No contact language—whether a stabilized pidgin, a second-language version of the lexifier, or a nascent creole—could have survived in its entirety through such vicissitudes. Indeed the only likely outcome is that the few who might have spoken some prior contact language would have been forced to abandon it and adopt the same formless mode of ad hoc communication that new immigrants had to use in order to communicate with one another.

The Sranan and Saramaccan Lexicons

The norm for creoles is that a large majority of words will be drawn from a single language. While in both languages English words are commonest, in Sranan there is an unusually high percentage of words from other languages, and in Saramaccan English does not even have a majority, only a plurality. The African component in Saramaccan is very large; Price (1976) estimated it at 50 percent, which may be an exaggeration, and surprisingly there seems not to have been any subsequent estimate. In both languages, words are drawn from English, Dutch, Portuguese, several African languages, and several Amerindian languages. Remarkable too is the high number of synonyms. Normally creoles reduce rather than increase synonymy; for example, a single verb often subsumes all the functions of "speak," "say," "tell," "talk," and so on. Why would a single language have had two or more words for the same thing, sometimes even two words from the same language? The most cursory examination of the two lexicons indicates that something quite uncharacteristic of creoles must have been going on.

Typically it is pidgins rather than creoles that are macaronic. This is a natural consequence of the fact that pidgins are neither a botched attempt to learn a dominant language nor an attempt to assert the speakers' identity and differentiate them from their masters—probably the two most popular theories of what pidgins are. Rather they are a negotiation between speakers of mutually unintelligible languages who are simply seeking ways to communicate with one another. The process was well described to me by someone who had actually been through it: "So we use the Hawaiian and Chinese together in one sentence, see? And they ask me if that's a Hawaiian word, I said no, maybe that's a Japanese word we put it in, to make a sentence with a Hawaiian word. And the Chinese the same way too, in order to make a sentence for them to understand you" (Rachel Kupepe, cited Bickerton 1981: 11). Substitute

Fongbe, Kikongo, and Portuguese for Hawaiian, Chinese, and Japanese, and you probably have a pretty good picture of what was happening in Suriname prior to 1690. In Hawaii the continued presence of English ensured that this state of affairs could not continue. But in Suriname English was removed, so it did continue.

There are several studies that use the lexicons of one or both languages to analyze phonology or determine the provenance of individual words (e.g., Daeleman 1972; Smith 1987, 2008; Kramer 2002; Good 2009), and there is one book-length study that deals with early Sranan vocabulary only and does not explore the issues examined here (Braun 2009). There has not, to the best of my knowledge, been a single study directly comparing the Saramaccan and Sranan lexicons. Any lexical studies that mention differences between these languages deal only with the English-Portuguese dimension. This is all that supposedly differentiates Saramaccan from Sranan; it results from the fact that a sizable percentage of the original Saramaccan speakers came from plantations owned by Portuguese Jewish refugees. (For a brief history, see Goodman 1987.) But in fact English-Portuguese alternatives yield under a third of the lexical differences between the two languages.

If two languages have a recent (less than four hundred years ago) common origin and hence something close to a common grammar, one would not expect to find an almost 50 percent difference between their vocabularies. Still less would one expect a widespread difference in the actual grammatical items through which that grammar is expressed. But as Table 8.1 shows, these differences exist and are widespread. In some cases they arise through choice of words not from different languages but from the same language, English. To express imperfective aspect, Saramaccan chose *ta*, from "stand," while Sranan chose *de*, from "there" (Winford and Migge 2007), and for "where," Saramaccan used a reduction of the word "side," while Sranan used a reduction of the word "place." Since grammatical functions can hardly exist in the absence of function words to express them, one can only assume that each language made, from the different possibilities offered by a highly variable pidgin, different choices for the markers of many grammatical functions.

Similarly if Sranan and Saramaccan had come from a single already developed language, one would not expect to find that more than half the days of the week and all the months of the year were completely

Creolization

Table 8.1. Grammatical items in Sranan and Saramaccan

	Sranan	Saramaccan
Definite article (sing.)	(d)a	di
Definite article (plural)	den	dee
Imperfective aspect	(d)e	ta
Anterior tense	ben	bi
WH-words:		
who	suma	ambe
what	faa, san	andi
when	oten	na unten
where	(o)pe	ka, naase
which	sortu	un
why	sanede	andimbei
Pronouns:		
you (sing.)	yu	i
we	unu	u
Prepositions:		
in	ini	a
along	psa	langalanga
inside	ini(sei)	dendu
under	ondro	basu
Conjunctions:		
and	e, nanga	ku
if	efu	ee

different in the two languages. The four days from Monday through Thursday—*munde, tudewroko, dridewroko, fodewroko* in Sranan—are *fodaka, feifidaka, pikisaba, gaansaba* in Saramaccan.

Saramaccan months are particularly interesting, since several of these reflect activities or states of affairs peculiar to a rain-forest society; in other words they must have postdated escape from the plantation system: *baimatuliba,* July ("sweep-forest month"); *hondimaliba,* June ("hunter's month"); *sebitaaliba,* May ("leech month"), *tanvuwataliba,* August ("wait-for-water month"). From this, and from the fact that Sranan adopted its months directly from the Dutch months, one can only conclude that the original Suriname pidgin did not have names for any of the months and that the Saramaccan names developed only after 1690 in the interior rain forest.

Creolization 247

An established language would surely also agree on the terms it adopted for the four cardinal points. But once again Sranan and Saramaccan have very different sets: Sranan *noordsei, zuidsei, oostsei, westsei* (compound words that combine the Dutch words for the cardinal points with "side") as against Saramaccan *basuse, libase, sonukumutu, sonugo* (literally "bottom side," "topside," "sun-come-out," and "sun-go"). The word *liba* by itself gives us considerable insight into the processes that gave birth to the Suriname creoles. In Sranan, *liba* means "river," which is a straight English survival. In Saramaccan, "river" is *lio*, from Portuguese *rio*; *liba* means "month," "moon," "on top of," and "sky," and also part of the compound *libabuka bia*, literally "on-top-of-mouth beard" = "moustache." (Note that *libase* becomes "south," because the south side is where high land and the upper reaches of Suriname's rivers lie, while *basuse* indicates the northern lowlands.) Since Portuguese has separate words for all meanings of *liba* (*lua* "moon"; *mes* "month"; *ceu* "sky" and so on) but no form from which *liba* can be derived, the only plausible source is Spanish *arriba*, "up," suggesting that at least some of the "Portuguese Jews" were in fact Spanish speakers. But if Saramaccan developed from a Portuguese creole or "Jew-tongo" (Smith 1987), how could its speakers be unacquainted with the Portuguese versions of such basic vocabulary items as "moon," "sky," or "month"?

Similar differences between the Sranan and Saramaccan vocabularies are found throughout the lexicon. The words for some of the commonest tropical fauna differ across the two languages, including the words for "ant," "anteater," "mosquito," "cayman," "vulture," "eagle," "heron," "crab," "rat," "duck," "lizard," "centipede," "chigger," and "fly." Relatively few of these differences involve Portuguese etyma, and in one case that does ("lizard": *kaluwa* in Saramaccan, *lagadisi* in Sranan) it is Sranan, not Saramaccan, that chooses the Portuguese etymon *lagarto*. Again contrary to expectation it is sometimes Saramaccan that preserves the English-derived word, for example, *hansi*, "ant," as against the Sranan *mira* (from Dutch *mier*). Similarly the words for many body parts differ; these include "ankle," "thigh," "elbow," "waist," "forehead," "neck," "navel," "penis," "vagina," "anus," "groin," "skull," and "cheek."

Perhaps the most noteworthy feature of the two vocabularies is the high percentage of compound words that have different forms for the same meaning. Compound words are perhaps the most reliable sign of when the creole stage was reached. Such words typically fill the gaps in the pidgin vocabulary where there was no available word from any of the languages—English, Portuguese, Dutch, and an indeterminate number of African (not to mention Amerindian) languages—that were in theory available. Note too that these compounds often draw on two or even three different languages, such as *agofutu wojo*, "ankle," from *ago*, "knot" (source language unknown), *futu*, "foot," and *wojo*, "eye" (from Ptg. *olho*). This suggests that at least until the Sranan-Saramaccan split the overall Suriname vocabulary was extremely restricted in its semantic coverage but quite rich in its stock of common synonymous words drawn from a variety of languages—just the kind of thing that Rachel Kupepe's account of pidginization predicts. Although English is regarded as the main lexifier of both languages, only about seven hundred English roots survive in the two languages (Smith 2006).

To ensure that the analysis was not skewed by particular registers or semantic classes, a sample of one thousand Saramaccan words was compared with their Sranan equivalents. For convenience, online dictionaries of the two languages were used. For Saramaccan, the source was the Saramaccan-English dictionary produced by the Summer Institute of Linguistics (http://www.sil.org/americas/suriname/Saramaccan/Saramaccan.html). For Sranan, the sources were the Sranan-English dictionary also produced by the Summer Institute of Linguistics (http://www.sil.org/americas/suriname/sranan/Sranan.html) and the *Wortubuku ini Sranan Tongo* edited by John Wilner (http://www.sil.org/americas/suriname/sranan /sranan-English%20Dict.PDF). The sample was composed by starting from A and eliminating all proper names, all Saramaccan words for which no Sranan equivalents were listed, and any terms unlikely to have been in use in the seventeenth century. Words that were phonologically different but derived from the same etymon were treated as the same. Of the one thousand total, 521 words were the same in the two languages, while 479 were different.

It is hardly likely that two languages derived from a single developed language would show a difference in their vocabulary of nearly 50 percent after a little over three hundred years of separation. Less than a third of this difference can be attributed to the Portuguese element in the vocabulary. Most interesting were the compound words. Compound words normally appear in creoles far more frequently than they do in pidgins. Pidgin speakers can always revert to their native language for words; for creole speakers, creole is their native language, and their control of others is limited, if it exists at all. Pidgins are notoriously short on words, so one of the first things children must do is expand their vocabulary, and compounding offers a quick and transparent way of doing this.

Out of 105 Saramaccan compounds, only sixteen were found in Sranan. This gives a far higher difference between the vocabularies than the one thousand count. Given the fact that the richest period for compound formation is that which immediately follows the start of creolization, this sample, though smaller, is perhaps even more significant. Equally significant is the fact that well under half of the exclusively Saramaccan forms—thirty-four out of eighty-six—owed their difference to the fact that at least one member of the compound came from Portuguese.

Since the vocabularies of the two languages have never been thoroughly compared, it is possible, albeit unlikely, that a more exhaustive study might alter these findings. However, Surinamese historical demography clearly predicts that an unstable, relatively underdeveloped and highly macaronic pidgin would have persisted until the 1690s, and that prediction is amply borne out by the two vocabularies. Consequently the two languages must have developed separately after 1690, and their grammatical similarities can have come about only through the operation of the same skeletal UG on the same variable, unstructured pidgin.

Creolization in Hawaii

Pidginization and creolization in Hawaii might seem at first sight to have taken place under circumstances quite different from those in the Caribbean. Apart from Portuguese and English, none of the languages in contact had previously been involved in these processes (or at least, not in the same way; Chinese Pidgin English was a contact language for traders, not plantation workers, and it never creolized). Sugar workers were indentured laborers, not slaves. The area was not originally a colony but at least nominally an independent state with its own indigenous language. Slave children in the Caribbean had no formal education, while most if not all children in Hawaii did, and from about 1890 on, that education was (or at least, was supposed to be) in English. Yet the linguistic results were so similar to those in the Caribbean that many creolists (Goodman 1985; Holm 1986, 1988; McWhorter 1994), following Reinecke (1969), were convinced, prior to the publication of Roberts (1998), that HCE had been created by some form of diffusion from the Atlantic creoles.

There are differences, and there are differences that make a difference. The two should not be confused. Nobody has explained what linguistic differences would be made by the difference in status between slave and indentured laborer. Indeed that status was more similar than might be supposed: indentured workers in Hawaii could be fined or jailed for breaking their contract, they were quite often beaten and occasionally killed, and a sugar worker's strike in Kauai in the 1920s (Reinecke 1997) claimed twenty victims—as many as died in a number of minor slave revolts. The presence of an indigenous language complicated matters and delayed the onset of creolization but had no other observable effects. The effects of schooling can be exaggerated. An important distinction made by many in the study of second-language acquisition is that between input and intake; "language acquisition entails not just linguistic input but *comprehensible* linguistic input" (Long, 1996: 414). That relatively little school-English input was transformed into intake is suggested by remarks such as those of a contemporary observer who noted, "English was stopped like the study of Greek when the lessons were over" (*Pacific Commercial Advertiser* 1888). If it had been so transformed, that would merely make it all the more remarkable that *any* creole features appeared in HCE.

The major factors that conspired to produce creole languages were exactly those described at the beginning of this chapter. The plantation cycle described there is precisely mirrored in Hawaii's history. From the 1830s to 1876 a small sugar industry existed, corresponding to the homesteading phase in Caribbean colonies. During that period a stable and expanding pidgin was developing, but one based on Hawaiian rather than English (Roberts 1995). That pidgin was completely destabilized by the large-scale introduction of foreign labor that began quite abruptly in the late 1870s and continued unabated until 1930. During that period workers were brought together from China, Japan, Korea, various Pacific islands, the Philippines, Puerto Rico, Portugal, Norway, Germany, and Russia, the vast majority of whom could communicate only in a macaronic, highly unstable pidgin. In other words, all of the factors responsible for creolization in the Caribbean were present in Hawaii.

Thus none of the differences between conditions in Hawaii and those in the Caribbean invalidates the use of Hawaii as a model for understanding pidginization and creolization as general processes. The reasons for choosing it as a model are obvious: it is the only plantation creole for which abundant written materials are available for every stage throughout the pidgin-to-creole cycle. For all other creoles, there are no substantive records prior to a time when the process of creolization was well under way if not already completed, and far fewer than one would wish for subsequent periods. Hence it is vital that the Hawaii data be described and analyzed with the greatest thoroughness and care.

The scholar most responsible for our knowledge of that data is undoubtedly Sarah Roberts (1995, 1998, 2004), whose patient and thorough archival research uncovered massive amounts of information never tapped by previous researchers. Unfortunately Roberts's subsequent analysis of the data is seriously flawed in at least three respects. She overestimates the extent to which the first generation of creole speakers acquired parental languages, and she attributes to the substratum features that appear widely in other creoles, even though HCE contains no substrate languages found elsewhere. (Portuguese, a substrate in HCE, was a lexifier in Papiamentu!) But the most crucial flaw, and possibly the cause of the other two, is an extraordinary decision about who counts as a creole speaker and who does not.

The Status of Generation G2

In earlier work, Roberts (1998) showed convincingly that HCE originated among locally born children rather than immigrant adults, a finding that decisively disposed of the diffusionist theory of creole origins (e.g., Goodman 1985) as far as Hawaii was concerned.

However, in her dissertation Roberts (2004) divided the Hawaii population into three generations: G1 consisted of immigrants to Hawaii, G2 of their children, and G3 of the children of G2. Roberts states that "the second generation *took shape* in the 1880s and 1890s and *was joined* around 1905 by the third generation" (203, emphasis added). The italicized words defy analysis: Does she mean they were born then, or reached maturity then, or attained some indeterminate stage in between? Subsequently her table 4.4 merely compounds the mystery, since the respective start dates for both G2 and G3 across the four major immigrant ethnic groups (Chinese, Portuguese, Japanese, and Filipino) cover a span of thirty years, during which the overall linguistic picture drastically changed. Moreover these spans overlap, so that Filipino G2 begins twelve years after the start of Chinese G3! This fact alone (plus ignoring the role played by the indigenous population) already renders the G1–3 partition of doubtful utility for any linguistic purpose.

But this is of relatively little importance as compared with an unprecedented move that Roberts (2004: 220) now makes: "In the discussion which follows... I will use 'P/C' as a general label for HPE [Hawaii Pidgin English] and HCE—particularly to refer to proto-HCE at a time when it was not clearly distinguished from HPE." Consequently speakers of G1 are hypothesized as speaking HPE, G3 as speaking HCE, and the intermediate generation G2 as speaking "P/C."

There is no motivation for this in the data. Roberts never defines "proto-HCE" or "P/C" in linguistic terms. She does not cite any examples of either nor provide information about any structural feature(s) that might distinguish them. In fact her entire corpus does not contain a single example that cannot, on purely linguistic grounds, and without knowing the identity of the speaker, be unambiguously assigned to HPE or HCE. (See Table 8.2 for six major distinguishing features.) For all the early examples of HCE for which any information on the age of the speaker is given, he or she is (with only a single exception) defined as "a child" or "an adolescent" or something equivalent.

Compare the following typical examples of early HCE with the HPE examples (1)–(7):

14. That fella bin think he more smarter than me, but I never 'fraid for that thing he bin tell (1909).
15. You bin say go up on roof and paint him but I no hear you say come down (1913).

Table 8.2. Grammatical features in HPE and HCE

	HPE	HCE
All subcategorized arguments expressed	often not	always
Complex sentences	very rare	frequent
Embedded sentences	no	yes
Systematic TMA marking	no	yes
Canonical SVO ordering	no	yes
Regular determiner system	no	yes

16. The Indian he bin get one gun; he bin shoot one cowboy (1916).
17. William punch my father and blood bin come outside (1916).
18. Sometime my father take me for I go look the horse race with him (1916).
19. More better for I write the answer (1917).

Such examples are little different from contemporary HCE. HPE differs from HCE along the dimensions shown in Table 8.2, where six major features are shown to be consistently present in HCE but consistently absent from HPE. Therefore the arbitrary introduction of a nonexistent intermediate cannot be justified on linguistic grounds.

Roberts's decision to create some kind of stepping-stone between HPE and HCE seems to depend not on data but on the belief (contra the consensus among professionals in the field of bilingualism already noted) that a child can have only one native language. Her reasoning goes as follows: if children acquired a parental language, they had to have acquired it before they encountered HPE; therefore whatever resulted from that encounter, however creole-like in all linguistic respects, could not be their native language. Since a creole has traditionally been defined, in contrast to a pidgin, as a language with native speakers, it would then follow that whatever they spoke could not be a creole. Note that this argument depends crucially on the claim that children in the early twentieth century in Hawaii did indeed achieve native mastery of their parental languages. But did they?

G2 and Parental Languages

Roberts (2004) cites numerous extracts from Hawaii residents' recollections of their childhood. The first selection (231) purports to show that parents used only or mainly their native languages when speaking with their children. But the really important issue here is whether children developed sufficient mastery of their parents' language to be regarded as native speakers of that language, with the full range of knowledge that the term *native speaker* suggests to professionals and lay persons alike. For if they did not, their viability as vectors of superstrate transfer is severely compromised.

There is an extensive literature on child bilingualism, and a wide consensus of scholars (e.g., Petersen 1988; Meisel 1989; Dopke 1998; Yip and Matthews 2000; Genesee and Nicoladis 2009) agrees that "language dominance is seen as the major determinant of transfer" (Yip and Matthews 2000: 193). In other words, "balanced bilinguals," children with roughly equal competence in both languages, seldom make transfers between them, while "dominant bilinguals," children more competent in one language than the other, do make transfers, but only from the dominant to the nondominant language. One might imagine that in creolization an established, structured language would be dominant over a pidgin. It is, but only in a social or political sense. What "dominance" means in a linguistic context is not the social status or degree of development a language has reached but the degree to which each language is used and understood by a particular individual. Unless children knew their parents' native language thoroughly, their parental language is not dominant and chances of substrate transfer are very low.

There exists a startling disconnect between Roberts's claims (that children spoke parental languages natively but HCE nonnatively) and the evidence she herself presents, which indicates the reverse. Roberts (2004: 233) produces seven extracts to show that "children had learned their AL [ancestral language] *to some extent*" (emphasis added). The extent to which they had actually learned it is clearly demonstrated by the extracts themselves: "I can't talk much Hawaiian"; "I do not know enough Chinese"; "I can understand and speak only a childish Chinese language"; "We make such a mess of [speaking Portuguese] that we are laughed at"; "I never learned to speak the Portuguese language except a few baby words"; "A Japanese boy seldom speaks Japanese"; "I cannot speak Japanese very well—at home I spoke pigeon [*sic*] Japanese."

Roberts (2004: 233) speculates that "they may have originally had native competence in their AL" but does not produce any supporting evidence. What makes her speculation particularly implausible is that these recollections "were written by junior high, high school and college students" (227), that is, by persons only a few years away from the experiences they describe, which hardly gave time for memories (or languages!) to have been lost or distorted. Her subsequent claim that children did not encounter pidgin until they entered school is supported by only three extracts (234; all three are from Japanese) and is refuted by every one of her other examples.

Altogether Roberts (2004) presents a grand total of twenty-nine autobiographical extracts, drawn from a database of 731 linguistic autobiographies. If these extracts were the best evidence for her claims that she could provide, one can only wonder what the remaining 702 would look like. Even if we grant bilingual status to G2, G2 speakers were clearly creole-dominant rather than AL-dominant. We have their own utterances as testimony to their creole fluency and their own testimony as to their minimal mastery of their AL. Recall Siegel's claim that tensed purposive clauses in HCE derived from Portuguese subjunctives. What is the likelihood that children who "make such a mess" of speaking Portuguese that they arouse laughter or who can speak only "a few baby words" will control Portuguese subjunctives?

The only conclusion that can be drawn from the Hawaiian data is that children created HCE in a single generation and that they did not do so (and could not have done so) by strip-mining their AL for structures that the pidgin lacked, because they did not know their AL well enough. Therefore they must have used the only other possible source of structural knowledge: an innate and universal biological program for language, adding to it whatever material they could mine from HPE to fill just some of the underspecifications of that program.

Creoles and the Universalist Case

Chapters 6 and 7 showed that both variation and change in language and the primary acquisition of language are equally and fully consistent with the theory of language presented here. But neither provides arguments that rule out alternative theories quite as explicitly as creolization does.

While some generativists have been quick to accept any theory that referenced UG, and some attempts have been made (e.g., Roberts 1999; see also commentaries on Bickerton 1984b) to identify creole features with default settings of linguistic parameters, no generative linguist has tried to systematically incorporate creole languages into a general theory of language change and development. Nor is there any general agreement among generativists as to the precise role that creole origins should play in generative theory. Consequently the few generativists who have specialized in creoles have felt free to adopt diametrically opposite views on how creoles originated, even if they happen to specialize in the same creole. For instance, though both write almost exclusively on Haitian, DeGraff (2007) claims that creoles develop from their main lexifiers by normal processes of language change, whereas Lefebvre (1999) claims that creoles preserve the structures of substratum grammars while relexifying them (expressing those structures through lexical forms from the dominant language). Other approaches to language, including empiricist approaches, have had even less to say about creoles than generativists. In short, no previous general theory of language has been able to produce a coherent explanation of how and why creoles originated and how they obtained the large set of features that they share.

As this chapter has shown, creoles develop when social, historical, and demographic forces conspire to prevent children from fully accessing any preexisting language. Under those circumstances children continue to do what they are programmed to do: they deploy the algorithms for structure building supplied by UG, but here, instead of simply adopting some set of strategies for underspecification repair that are used by an already existing language, they pick up fragmented materials from other languages and repair the gaps by using the simplest and most straightforward strategies that those materials permit. There is nothing exceptional about this. There is little if anything in creoles that cannot be found elsewhere in language and "normal" language change (although only in creoles do large numbers of the relevant phenomena appear simultaneously). There is only what, if all linguistic input suffers from "poverty of the stimulus," one can only describe as "pauperization of the stimulus"—reduction in quality or quantity of input well below levels found elsewhere.

The previous three chapters may seem to have taken us far from our original goal. But if Darwin's inspired guess—that humans' "powers" came from "use of a highly-developed language"—is correct, it is essential that we know exactly what that language was, as well as how and why it developed as it did. In order to demonstrate that the account of language given here is correct, or at least more nearly correct than previous accounts, it is necessary to show that it gives a better explanation of the data than alternative theories can offer, particularly in the fields of change, acquisition, and creolization. That done, we can return to Wallace's problem and consider a solution that, among other things, should relegate to history the decades-long debate between nativists and empiricists over the nature of language.

CHAPTER 9

Homo Sapiens Loquens

Perhaps the most reassuring thing I found out in the course of writing this book was that in trying to explain how humans evolved, almost everybody was right about something. Few of the factors that were mentioned in the literature didn't fit into the story somewhere.

That was the plus side. The minus side was that, like the White Queen in *Alice in Wonderland,* far too many writers in the field could believe six impossible things before breakfast. They could believe that, unlike any other species, humans developed unique and highly specific powers without first undergoing some equally unique and highly specific interaction with their environment. They could believe that those powers grew and blossomed in some kind of neural vacuum, while the human brain just sat there passively, getting bigger, of course, but not developing any specialized mechanisms for implementing those powers. They could believe that apes were the best models for our immediate ancestors, even though we talk and apes don't; we cooperate and apes don't; and we produce a constant and seemingly infinite stream of new artifacts and new behaviors while apes go through the millennia without producing either. They could believe that "building blocks of language" (Lorenzo 2012: 289) are scattered across a vast range of species, awaiting magical assembly in some lucky ape's brain. They could believe that whatever was the immediate focus of their interest—language, cooperation, cognition—was largely divorced from the other foci, which might at best play a supporting role. They could believe that, whatever their focus, it had a single major cause.

If their focus was language, the cause could be hunting or tools, or social competitiveness, female choice, or infant care. If their focus was cooperation, the cause could be reciprocal altruism or kin selection, or the detection and punishment of cheaters. The usual suspects were rounded up and put on parade, and when the perpetrator couldn't be positively identified, the response was, "Well, it must be all of them." In a sense, of course, that was true. But you can't try a conspiracy case without first unraveling the plot.

The Three Rooms and the Escalator

In hindsight it seems obvious that anything as complex as the "higher powers" of humans could only have come about through a long and tortuous history in which each of a series of separate episodes precipitated the next. One crucial clue to the solution of Wallace's problem is one of the very things that make it seem so baffling—the immensity of the cognitive gap between humans and all other species. The fact that there are no species with cognitive capacities partway between chimpanzee and human is surely telling us something.

There is an immense variety of species on earth and they occupy every conceivable kind of niche. If it had been possible to find a niche that bestowed only some human powers, some species would surely have found one by now. This suggests two things. One is that all the "higher powers"—language, cognition, co-operation, self-consciousness—are intimately interconnected in some way. The other is that a species intermediate in its powers between apes and humans is not possible. There is no mezzanine in nature's mansion. If you get on the escalator, you go straight up to the next floor.

But we have to bear in mind that the escalator could not have been the co-evolutionary spiral that writers on evolution so often invoke. Until less than a couple of hundred thousand years ago, the tools and behaviors of modern human ancestors differed from those of apes much less than they differed from those of modern humans. To find an appropriate image we should think of a series of rooms that you have to pass through to get to the escalator. There is only one door to that series of rooms. If you can't get through that door, you'll never get on the escalator.

The first room is the confrontational scavenging niche. There, humans acquired the basis for cooperation: unless everyone, kin and non-kin alike, combined to secure a carcass, nobody would eat any of it.

There, even more crucially, they acquired displaced reference. Acquiring displaced reference is one way, perhaps the only way, to get into the second room. At least it is the only way that has emerged, over billions of years of evolution, to get out of the straitjacket of standard animal communication—to leave the here-and-now for the limitless expanses of time, distance, and imagination.

In the second room is symbolism. This is the one room about which we can't be totally certain. In all rooms but this, the following rule applies: you can't just stay in that room and you can't go anywhere but the next room. But is it inevitable that having gotten displaced reference, you must go on to symbolism? Clearly not, since bees and ants didn't. But the effects of displacement on organisms with minuscule brains must surely be different in brains that are orders of magnitude bigger, that can hold finely-dissected descriptions of the world, and that constantly engage in something unknown to ants and bees, rich and varied patterns of social interaction between highly individuated animals. Besides, we know the result—we have unlimited symbolism, bees and ants don't—and I know of no equally likely route through which symbolism could have been acquired.

Once symbolism is up and running and being used in probably still crude modes of communication and thought, entry into the third room cannot be avoided. This is where the brain plays its part. One of the oddest things about the literature on human evolution is the rarity with which the brain's power of self-organization is referred to. That power is hardly news; it has been the object of comparative study for the last few decades, and is one of the evolutionary facts that can be seen as inescapable once you start thinking about it. Brains must have self-organized and re-self-organized innumerable times over the last half-billion years, as species after species increased and/or changed its sensory equipment. If brains hadn't changed so as to process new kinds of information, the species concerned would have had radically incomplete notions of what the world around them was like, and would quickly have gone extinct.

We know too that the brain stereotypes and automates physical actions like throwing or picking stuff up, so that these actions can be performed more quickly and efficiently. Why would it not do exactly the same for mental actions like thinking and uttering sentences? In any case the second of these requires muscular as well as neural activity. For the brain it's all one, mental, physical, mental-and-physical—all are performed in the same code of electrochemical discharges, patterns of firing and non-firing neurons. It is, or should be, unthinkable that a machine that can time the launch-window of a throw down to the millisecond level should have to fumble and stumble over arranging groups of words or concepts as if each time was the first time it had ever done this.

But reorganization as extensive as the human brain now had to undertake is a time-consuming business. A vast and dense forest of relatively local interconnections (between word and word, word and speech organs, word and concept, concept and concept) had to be constructed. Neural superhighways like the arcuate fasciculus had to link the more widely-spaced regions involved. Leaving the third room had to wait on completion of the process.

Once it was completed, humans had nowhere to go but up. They now had a machine to think with that was not just without parallel in the animal world—it did not have even remote competitors. But it was at this point that mind and language diverged.

It is easy to get confused here, because the relationship is shifting. The mechanism that the brain produced was originally neutral between thought and language. There was a single means for linking words and/or their associated concepts. When the brain was done reorganizing, this could be used with full efficiency for thought and with somewhat less efficiency for communication. Since then, thought itself has remained unchanged across the species (although what thought could construct has increased exponentially).

Once on the escalator, however, humans started tinkering with what we can now truly call language. It was while riding up the escalator, that is to say over the last 150 millennia, that we began trying in many different ways, more and more ways as the human diaspora proceeded, to make up for the many underspecifications of the brain's structure-building mechanism. We produced so many ways, so many "improvements" on the basic plan that the plan itself virtually vanished from sight, and it became easy for many people to believe that there was no such plan, that the brain had never developed any task-specific mechanisms for language. In a sense they were right; technically speaking, the mechanisms weren't *for* language. They weren't developed *for* anything, except to make the brain's tasks lighter and help conserve its energy.

What provoked their birth was prehumans' invention of displaced signs, a process the results of which the brain had to do something about. But the use they were put to thereafter was a matter of indifference for the brain. Both thinking and speaking involved arranging series of neural events in particular orders.

When humans began work on making language easier for the hearer to process, the picture changed. The structure-building mechanisms themselves could not change; they were a fixed part of the human phenotype. They could, however, acquire a learned component, and in Chapter 6 we got some idea of how far that component could ramify. The cultural and biological elements of language grew so closely intertwined that sorting them out looked not just a difficult but perhaps an unnecessary or even misguided task. It was so easy to say, "It's all biological" or "It's all cultural."

The foregoing provides at least the outline of a solution to Wallace's problem. Mind and language were more than nature needed because only the first of these four stages resulted from particular selective pressures operating specifically on human ancestors. The second stage grew from the tendency of displaced-reference units to refer in ever more general and abstract ways until they became, in effect, labels for mental concepts. The third stage consisted of purely brain-internal operations responding to the unusual phenomena that the first two stages had presented to the brain rather than on any selective pressure exclusive to hominids. Only the fourth stage began to produce natural language as we know it, and by that time the earlier stages had generated mechanisms whose potential power bore no relation to the ecological needs of humans.

"Continued Use" and the Joyce Factor

But this still leaves a lot to be explained. Wouldn't it be easy to imagine primitive humans who could talk as fluently as you or I but whose intelligence was no different from that of a chimpanzee? If there wasn't some additional factor involved, something quite other than language, how could cognition as well as communication have changed so radically in the past few million years?

Here we need two more clues. One lies in Darwin's original conjecture as to the possible function of language cited in Chapter 1. Darwin was usually careful in his wording, so we should pay equally careful attention to it. What he wrote was this: "If it be maintained that certain powers, such as abstraction, self-consciousness etc., are peculiar to man, it may well be that these are the incidental results of other highly-advanced intellectual faculties; and these again are mainly the result of the *continued use* of a highly-developed language" (1871: 103, emphasis added). The other comes from a much more unexpected direction and is what we might call the Joyce factor. This will remind us that "continued use" doesn't mean just "conscious or intentional use."

First, I want to dispose of a particular misapprehension that I feel partly responsible for, having not always followed Darwin's example in precision of wording. It would be all too easy to summarize the take-home message of this book as "Advanced human cognition results from language." That is inaccurate even as shorthand. "Without language, advanced human cognition could not have existed" is better but still inadequate. It could be rephrased as "Human ancestors began to communicate with displaced reference, and that was what triggered the processes that eventually led to advanced cognition." But that, if more accurate, is unlikely to catch on as a slogan.

In their seminal article, Penn et al. (2008) consider three possible versions of the "language causes cognition" claim. The first would claim that "natural language sentences are responsible for the disparity between human and nonhuman cognition" (121). This is certainly not what is claimed here. We don't think in sentences, we think in spike trains, which as noted above are neutral as between thought and language. The second claims that "some aspect of our 'language faculty'" (121) is responsible for the cognitive gap. This merely restates the general "language causes cognition" claim in a way that is quite meaningless unless specific candidate aspects are proposed. (Unlikely candidates discussed by Penn et al. include recursion and logical form.) The third attributes human-nonhuman differences to "the communicative and/or cognitive function of language" (121). But here the distance between Penn at al.'s imaginings of what language might do for thought and what is proposed in this book becomes clearly apparent.

They state, "It is quite difficult to imagine how communicating in hierarchically structured sentences would be of any use without the ability to entertain hierarchically structured thoughts" (Penn et al. 2008:

123). It is much harder to see how either could exist without the other. The notions that thought and language are clearly separable entities, each with its distinct infrastructure, and that language merely "expresses thoughts" seem very natural ones. Certainly it is extremely difficult to root them out (see Bickerton 1992b and the response by Newell 1992). But what has been proposed here is that language and cognition (at least those aspects of cognition that are unique to humans) grew from a common source and are based upon the same foundations.

Those foundations are twofold: the creation of symbols that can be consciously manipulated (the ultimate consequence of exploiting the confrontational-scavenging niche) plus the means to manipulate those symbols *and therefore also the concepts they represent* in order to produce novel configurations (derived from the neural reorganization necessary to accommodate the utterance of words and propositions). Thus it is not a case of language begetting human cognition but rather of processes that in and of themselves were neither specifically cognitive nor specifically linguistic giving rise to distinctively human cognition and language at (approximately) the same time.

To Penn et al. (2008), the difference between human and nonhuman cognition consists in human ability (and nonhuman inability) to reason about abstract, higher-order relationships where the objects or categories compared can be distinguished only by their roles or functions (lovers, mothers, tools, etc.) rather than on the basis of perceptual or statistical properties that they share. They claim that the capacity to do this does not reduce to any other kind of cognitive operation and consequently find it necessary to hypothesize a specialized cognitive process, additional to the rest of the human linguistic-cognitive armamentarium, one they describe as "relational reinterpretation."

Grant their claim that reasoning about higher-order relationships does not reduce to any other single and distinct operation. Even so, hypothesizing mental mechanisms over and above all those that humans are already known to have (ones that require clumsy neologisms too) is vulnerable to Occam's razor. Moreover proposing any such mechanism commits one, sooner or later, to having to explain how it evolved. It is not enough to claim, as Penn et al. (2008) do, that it would have increased individual fitness. So would lots of things—eyes in the back of one's head, a sense that would alert one to venous obstructions—but such things are simply not evolvable. No variation that natural selection could work with nor any selective pressure to which the mechanism would serve as a response is apparent in any of these cases.

Moreover we still have to consider the full implications of Darwin's wording—that it was not so much the inherent properties of language that gave rise to uniquely human capacities as its *continued use*. It was continued use—not, *pace* Darwin, of language per se, but of the basic symbolic and structural materials from which language would grow—that allowed what was in effect an entirely different form of cognition.

So far, in thinking about human thinking, the latter has been presented as typically a proactive, goal-directed process. I have focused on things like the complex trains of thought that, as Darwin pointed out, are difficult to construct without invoking words (even if these remain unuttered). Such trains, of course, constitute the crowning glory of human cognition. They are what has enabled us to make a potentially infinite number of discoveries about the nature of things, as well as a multitude of inventions that, we hope, improve the quality of our niche. But with respect to quantity, these conscious and intentional thought-trains amount to far less than the tip of the iceberg of mental activity, constituting probably less than .001 percent of it. It is the nature of the remainder of that activity—what you might call the dark matter of cognition—that we should now examine.

The Joyce Factor

"He heard then a warm heavy sigh, softer, as she turned over and the loose brass quoits of the bedstead jingled. Must get those settled really. Pity. All the way from Gibraltar. Forgotten any little Spanish she knew. Wonder what her father gave for it. Old style. Oh yes, of course. Bought it at the governor's auction. Got a short knock. Hard as nails at a bargain, old Tweedy. Yes sir. At Plevna that was. I rose from the ranks, sir, and I'm proud of it. Still he had brains enough to make that corner in stamps. Now that was farseeing."

All but the first sentence describe thoughts that pass through the mind of Leopold Bloom, a central character in *Ulysses* (Joyce 1960), as, having brought his wife breakfast in bed, he finds her still asleep and carefully withdraws. At the sound made by the brass quoits as she turns, his mind leaps to the bed and its provenance, with passing reference to his wife (who lived in Gibraltar as a child), and then switches to the bed's buyer, his father-in-law, remembering remarks that were typical of him and hedging an initial low estimate of the man's intelligence.

This famous steam-of-consciousness technique, of which Joyce was the chief developer, is in words (as it has to be, to be accessible to the reader). Moreover fiction has to be neater and tidier than life. Yet within these limitations Joyce's technique still forms a vivid portrayal of the kind of mental activity all of us experience on a daily basis, the random chatter that is anathema to mystics and mathematicians alike, since it interferes with both meditation and the laser-beam focus required for solving highly abstract problems.

Stream of consciousness has been a topic for psychologists ever since James (1890) broached it. But emphasis has been on the "consciousness" rather than the "stream" and has focused on phenomena like "change blindness" (Simons and Rensink 2005; Simons and Ambinder 2005), whereby events of which observers should be cognizant (but aren't) are made to occur in their visual fields. Indeed many studies have dealt solely with aural or visual streams, ignoring the phenomenon Joyce tried to capture, which combines visual, oral, emotional, and other elements. Topics such as why some experiences are conscious and others not, or why we are self-conscious at all, will not be examined here—although the latter problem may well be overblown. (Given the power to construct continuous narratives with oneself as a major actor, it is hard to see how self-consciousness could be avoided.) Here I want to deal merely with the erratic jumping-about of the Joycean stream, its movement from topic to topic, constrained by chance associations rather than any focused behavior.

To start, let's ask whether other animals have a stream of consciousness or if this phenomenon too is limited to humans. Of what evidence there is, most suggests that other animals probably do have some kind of a stream of consciousness. The source of the phenomenon can only lie in the fact that an inactive, "resting" neuron can be fired by neurons that themselves have fired for some function in which the first neuron plays no part or that belong to a random chain of neurons formed through past occasions of (probably sometimes accidental) synchronous firing. Since from neurons to overall architecture, mammalian brains conform to a similar pattern, we have no reason to suppose that stream of consciousness is unique to human brains.

This position finds support in the nature of dreaming. There is a large and contentious literature on this subject (see *Behavioral and Brain Sciences* special issue on sleep and dreaming, December 2000), most of which involves issues (e.g., whether dreaming is limited to REM sleep or whether it has any adaptive function) that do not concern us here. But whatever the brain does with images so generated, it seems likeliest that the images themselves arise through what Fishbein (2000) describes as "chaotic cerebral activation"; the fact that the brain might subsequently put this phenomenon to some practical use, even use that might have been evolutionarily adaptive (Hobson 2009), in no way undermines this conclusion. Certainly anecdotal accounts of animals (especially dogs) that appear to be pursuing prey in their dreams are very frequent. If random activation is ultimately responsible for dreams, then dreaming and stream of consciousness are modes of the same phenomenon, differing only to the extent that censorship and editing are more relaxed in sleep. The default assumption should therefore be that other animals too have streams of consciousness. But the arrays of neurons excited would be different from those in the human version.

When a concept is evoked in the nonhuman version, it is unlikely to fire other concepts, since no mechanism for linking concepts with one another has yet developed. A concept is much likelier to trigger percepts associated with it or behavioral reactions to what it represents than it is to trigger another concept. When we hear a sleeping dog emit excited noises while rapidly moving its feet, we cannot even be sure it is dreaming, let alone know what the dream is about, but its behavior is surely consistent with the pursuit of dream rabbits. It therefore seems reasonable to assume that if other animals do have streams of consciousness, these are much more tightly linked with the animal's past experience than ours are.

So, quite early in the cycle of human development, the stream of consciousness would begin, at a glacial pace, to change. No one would know this, of course. But the changes that took place would involve an imperceptible increase in the frequency with which one concept fired another. And as the number of linked concepts increased, so would the chance that the stream of consciousness would successively bring together concepts that were not normally associated with one another.

An event of this order is where many new ideas come from, in particular those seminal flashes of thought that frequently result in new discoveries and inventions. We sometimes think of "Aha!" moments as results of conscious and deliberate mentation, and they sometimes are, but they can equally well arise quite unexpectedly. Most of us have probably experienced such moments, when the connection between two or more seemingly unconnected concepts suddenly appears to us so plainly that we ourselves are startled, yet at the same time puzzled as to why we, or somebody, "never thought of that before."

It seems highly possible that stream of consciousness was the original means through which advanced cognition emerged. The process does not at first look like anything one would classify as "intelligence." Just like syntax, it would have been as far below the level of consciousness and volition as breathing or digesting. But there is a difference between "unconscious thinking" and these more plainly physical processes. Thinking, conscious or unconscious, is concept linkage. But concepts, once evoked, unavoidably link to the words that represent them. And words have to be within reach of consciousness and volition, because speaking is (usually) a volitional act, and has to be so, since there's no knowing when you will need to say something. Consequently it may well have been inevitable that processes beginning as random, uncontrollable events should have become, over time, capable of being converted into conscious and volitional acts, aimed most likely at enhancing the thinker's prestige but having the incidental effect of contributing to the ever-growing edifice of human culture. To put it very simply, if crudely, it may have been through our involuntary thoughts that we became able to think voluntarily.

Note that the process would involve a combination of biological, developmental, and cultural factors. Obviously neither words nor concepts can be directly inherited. But overall brain architecture, numbers of neurons, wiring density, and major wiring pathways may all be under genetic control. Words and their accompanying concepts would be transmitted culturally across generations. New generations would then make small additions to that stock, ensuring that protolanguage would grow with the ratchet-like effect that is the hallmark of evolution generally.

However, the full flowering of the cognitive developments that this process made possible would have to await the development of algorithms for the concatenation of words and concepts. It is in this sense that the "continued use of language" that Darwin hypothesized—or, to be more precise, use of the elements that would ultimately give rise to language—was responsible for mental powers so far in excess of ecological needs.

Why We Need No Additional Capacities

The question now becomes, what if anything in typically human cognition has this account failed to explain. There seems to be a widespread feeling that the enormity of human cognition requires explanation by something at least equally exotic, some esoteric factor over and above the rather humble and familiar evolutionary factors described here. There must surely be some "representational rediscription" (Karmiloff-Snith 1992), "cognitive fluidity" (Mithen 1996), "double-scope blending" (Fauconnier and Turner 2002), or "relational reinterpretation" (Penn et al. 2008). The fact that a new neologism for the "X factor" seems to pop up about every five years makes one suspect that these terms, like Moliere's "dormitive principle," are little more than new names for what is being sought.

Let's return to the topics discussed in Penn et al. (2008): the types of problem that, in the authors' opinion, other animals can't solve but humans can. Note that the specialized mental operations subsequently discussed (matching to sample, discerning and following rules, understanding spatial, hierarchical, and causal relations, and making transitive inferences) may require the operator to discern relationships based on roles, functions, or other abstract criteria rather than on perceptually based features. But operations involving any degree of abstraction are inaccessible to any species that cannot represent abstractions.

Are there any other factors that distinguish problems that animals can solve from problems they can't? Penn et al. (2008) do not mention any, and it is unclear what such factors might be. Moreover the wide range of evidence presented by commentators on the article, and largely conceded by its authors, illustrate the effectiveness, in a wide range of mammals, birds, and even fish, of nonhuman reasoning capacities. Wherever perceptual features exist or there are clear statistical differences, those capacities function well. Problems arise solely where conclusions have to be based on abstractions—on roles such as parenthood or properties such as edibility, and even more so when relations between things rather than things themselves are being compared (e.g., is the relationship between lock and key the same as that between corkscrew and bottle, or between skin and banana?).

Accordingly, the null hypothesis is that nothing beyond the cognitive capacities of other animals and the series of processes described in this book was required for modern humans to achieve all the powers they now exhibit. The symbolic and syntactic mechanisms that those processes created, when combined with pre-existing cognitive capacities, acted like a supercharger, multiplying the force and scope of those capacities. If we are looking for evolutionary continuities beyond mere physical form, here is where to find them. Humans have no additional reasoning powers. They merely have a means to augment the powers that many of the more complex species already have.

Why that means works so well, giving the illusion that a whole host of new mental abilities has suddenly and inexplicably emerged in just one species, ours, is because it also creates what Mithen (1996) was trying to capture with his expression "cognitive fluidity." As we have seen in other species, from spiders and bees to scrub-jays, beavers, and bats, high levels of ability in niche-specific skills never spill over into general cognition. But one effect of the chain of processes described here was to provide a kind of common code for all forms of mental activity. In the words of Boeckx (2012: 498) it conferred "the ability to combine virtually any concept (from whatever [core] knowledge system) with any other concept (from the same or another knowledge system)," and thus "mixes conceptual apples and oranges in virtue of them all being word-like things." This is what creates the common illusion that some unique and incredibly powerful "general cognition" has somehow, without being rooted in any nameable evolutionary process or event, managed to develop in humans but in no other species.

As for specific continuities in cognition, we need look no further than the need to determine causation. This is only a part, although a vital one, in the general need of all sentient organisms to construct in their minds the most comprehensive and accurate model of the world that they can achieve. The more accurate such a model is, the longer and more productive the life of that model's maker. Consequently natural selection will favor anything that improves the accuracy of the model. To take an obvious case, if an animal can determine whether a particular ripple in tall grass is caused by a gust of wind or the stealthy approach of a predator, this will reduce its chance of a sudden and premature death.

Clearly, better and more complete models of the world deliver higher levels of evolutionary fitness. A perfect model would include the causation of everything. (Scientists spend their careers looking for bits of that model.) Wherever we observe a substantive consistency we assume it does not result from mere coincidence. If all unsupported objects fall, there must be a cause for this, whether you call it "gravity" or "curvature of space-time." And since we have words, we don't need a ready-made concept to label. We can make one up from those very words ("whatever it is that makes everything fall") and then attach to it whatever attributes we deem appropriate. Our capacity to infer causation is shared with many other species. The extension of that capacity over a wider range of phenomena, including more abstract phenomena like gravity, is all that creates a discontinuity here.

Some Consequences of These Proposals

Proposing a solution for Wallace's problem is not quite like a detective solving a crime. True, in both cases, any proposed solution has to run the gauntlet of a confirmatory procedure: a formal trial if a crime is involved, the prolonged scrutiny of colleagues, including experts in all the fields involved, if we are dealing with a theory. But in the former case, the original solution is upheld oftener than not. In the latter, things are very different. No one would be more shocked and surprised than its author if the theory advanced here were to be unanimously and unquestioningly accepted. In scientific inquiry things just don't play out that way. While I feel reasonably sure that, given all the considerations taken into account here, something along the general lines of this theory has to be correct, many of its details will surely require revision or replacement as they are further scrutinized.

However, the really significant difference lies in the sphere of consequences. When a crime is solved, that's that: case closed. But a theory of any human behavior, even if it should prove both comprehensive and accurate, inevitably raises more questions than it answers and provokes new research in a variety of areas. In the present case, there are implications for the further study of issues as remote from one another as the brain's construction of sentences and the causes and courses of linguistic change. Space precludes the examination of all these, but two are of sufficient general interest to merit a quick summary. I would like to close this book by briefly discussing the evolutionary probability of animals with human capacities and the decades-old debate between nativists and empiricists.

How Probable Were Humans?

According to Gould (1989), the emergence of humans was a wholly unpredictable and evolutionarily improbable event—one that, if the "tape of life" were to be replayed, would never recur. Similar opinions have been voiced by many biologists (e.g., Simpson 1964; Mayr 1995) and cosmologists (e.g., Tipler 1980). Even Conway Morris (2003), stout believer in evolutionary convergence and challenger of both Gould's interpretation of the Burgess Shale and his belief in the improbability of human intelligence, regards it as unlikely that similar phenomena can be expected on other worlds. This is not, however, the scenario that the present theory projects.

Recall that the most striking aspect of the situation that so confused Wallace is the complete absence of animals linguistically and cognitively intermediate between humans and the great apes. This gap cannot be explained away by the extinction of intermediate species in the human lineage, despite the frequency with which these extinctions are invoked by continuists. As noted earlier in this chapter, even though what triggered the growth of human capacities could have occurred two million years ago, the full flowering of those capacities came only with our own species. For several million years after they split from other apes, human ancestors were far closer to apes than to modern humans in their behavior, so the cognitive gap remains undiminished.

Moreover, the last common ancestor of apes and humans can only have been some fairly ordinary ape. If it is true that our lineage went from that ancestor to its human conclusion by a long series of gradual steps, then it is strange that not one of those steps was undertaken in any other lineage, primate or nonprimate. This could be the case only if the first step was an essential prelude to the rest, as in the "three-rooms-plus-escalator" scenario described earlier in this chapter. And the usual accounts of human evolution, which see gradual enhancement of a whole suite of different ape behaviors as its driving force, are unable to explain how improvement in any one of these areas could have been a necessary condition for improvement in others. Vague talk of coevolutionary spirals does not suffice here.

It is surely relevant that, in contrast with the consequent human isolation, there should be a number of species in widely separated lineages (primates: chimpanzees, bonobos, gorillas, orangutans; cetaceans: dolphins, whales, even sea lions; and birds: jays, crows, parrots) that appear to be at a roughly similar level of cognitive development, crowding outside the door of the first room, so to speak. It is as if they were held up by some barrier that, if they could surmount it, would automatically place them on the path to the escalator's ineluctably moving belt.

This book has suggested that the barrier could be breached by displaced communication, which in turn could be achieved through the ecological developments surveyed in Chapter 4. Whether or not this represents the only way the barrier can be breached I shall not attempt to determine. But even if it does, there is still a strong possibility that, on any planet that hosts life forms, some species that has reached the chimpanzee-dolphin-crow level of cognitive capacity will eventually adopt a niche similar to that occupied by ants, bees, and human ancestors on this planet. If it does, then other planets with "intelligent life" become perhaps unavoidable.

In the week these lines were written, Harvard astronomers, analyzing new data from the Kepler telescope, estimated that there might be as many as 17 billion Earth-size planets in the Milky Way alone, a sizable percentage of which would have orbits within a zone congenial to life (Cowen 2013). Earth, far from being the galactic anomaly many previously believed, is as ordinary a planet as the last common ancestor of apes and humans was an ordinary primate. To speculate further is premature, but these findings strongly suggest that the array of life forms on these planets may differ little in their cognitive spread from those found here, and that consequently "intelligent life," far from being a rare or even unique aberration, may have multiple loci throughout the universe.

Meanwhile, back on Earth, the data surveyed in this book raise the possibility that evolution, far from being the random process envisaged in the previous century, may be quite constrained in the courses it can take. That it may be narrowly determined by biological as well as by physical and chemical laws, with sharply limited options at each change point, is a possibility the exploration of which should constitute one of the most fascinating fields of twenty-first-century science.

Nativism versus Empiricism

As we have had occasion to note several times already, the seemingly irresolvable debate between nativists and empiricists has persisted for several decades, to the detriment of our understanding of human behavior. Both sides have paid lip service to the notion that inborn characteristics interact with culturally acquired ones in every area of human behavior, but when it comes to certain key areas, in particular syntax, both sides dig in their heels and insist—empiricists, that there is no task-specific, genetically transmitted apparatus for the area in question, and nativists, that almost all if not all behavior in that area is tightly controlled by biology.

If the thesis of this book is correct, the unusual persistence of this debate is due to the fact that nativists and empiricists were both partly right (although they were both also partly wrong). Nativists were right in supposing that syntax couldn't exist without a substantive innate component but wrong in supposing that this component included most if not all syntactic structures. Empiricists were right in supposing that there was too much variability in syntax for it all to be produced by any dedicated universal system but wrong in supposing that all of syntax could be acquired without a substantial innate component. In retrospect it seems bizarre that nobody, throughout this debate, proposed a principled and systematic distinction between those parts of syntax that were biologically given and those that had to be acquired through acculturation into one of the many thousands of speech communities.

But what you have just read includes, and to a considerable extend depends on, the first coherent theory of syntax that makes such a distinction. It is also worth noting that it is the first theory to have begun rather than finished from an evolutionary perspective. Previous theories have taken some model of syntax as a given and only later (if at all) asked how such a model could have evolved. This theory began by asking how anything like syntax could have evolved and only then considering what kind of syntax was evolvable and whether that kind might lie at the core of modern syntacticized languages.

Is it too much to hope that the solution offered here will finally resolve the debate, or at least consign it to history? Of course it is. Embattled scholars with honorable scars and years in the trenches are not about to surrender positions they have defended throughout their careers. Young scholars at the start of their careers are, naturally, another matter.

Conclusion

In any case, resolving the nativist-empiricist debate was not the goal of this book—though the fact that a resolution emerged naturally in the course of pursuing a quite different goal surely suggests that the pursuit itself was along the right lines. Whatever the fate of the solution proposed here, I hope at least to have convinced readers of three things:

1. Wallace's problem was a very real one, one that lies at the heart of any understanding of what humans are and why they are what they are.
2. Neither Wallace nor Darwin could have solved the problem, since the necessary knowledge, only a fraction of which has it been possible to cite here, required the amassing of over a century of work by thousands of dedicated scholars in a wide variety of fields.
3. Given that knowledge, we no longer need to treat the problem like an embarrassing family secret that, if openly discussed, might offer aid and comfort to creationists and advocates of intelligent design. Wallace's problem can be solved, and indeed can only be solved, by resort to the successive action of two normal evolutionary forces (in this case, natural selection, neural reorganization), together with the subsequent development of human culture. The first two factors, individually, were constantly present throughout half a billion years of evolution; their interaction, though more sporadic, was still frequent. Only once, however, did that interaction result from attempts to exchange information beyond the bounds of the here and now.

The growth of both language and mind began at that crucial choice point where, as in Robert Frost's poem, two roads diverged—not in a yellow wood but in the tall lion-colored grasses of climax savanna. One, well-trodden by previous hominids, led only to bone crunching, predator evasion, and subsistence on discarded scraps. The other led to a new niche where the fiercest predators had to be confronted but where there was also the promise of plenty. Our ancestors took the road less traveled by, and that has indeed made all the difference.

REFERENCES

ACKNOWLEDGMENTS

INDEX

References

Abe, Takuya, David Edward Bignell, and Msdshiko Higashi. 2000. *Termites: evolution, sociality, symbioses, ecology.* New York: Springer.

Abney, Stephen Paul. 1987. The English noun phase in sentential aspect. Ph.D. diss., Massachusetts Institute of Technology.

Aboh, Enoch O. 2009. Competition and selection: That's all. In E. O. Aboh and N. Smith, eds., *Complex processes in new languages,* 317–344. Amsterdam: Benjamins.

Aboh, Enoch O., and Umberto Ansaldo. 2006. The role of typology in language creation. In Umberto Ansaldo, Stephen Matthews, and Lisa Lim, eds., *Deconstructing creole,* 39–66. Amsterdam: Benjamins.

Adams, M. D., et al. 2000. The genome sequence of *Drosophila melanogaster. Science* 287 (March 24): 2185–2195.

Adone, Dany. 1994. *The acquisition of Mauritian Creole.* Amsterdam: Benjamins.

———. 2012. *The acquisition of creole languages.* Cambridge, U.K.: Cambridge University Press.

Adone, Dany, and A. Vainikka. 1999. Acquisition of WH-questions in Mauritian Creole. In Michel Degraff, ed., *Language creation and language change,* 75–94. Cambridge, Mass.: MIT Press.

Aguilera, Mariela. 2011. Animal concepts: Psychological theories, pluralism and division of labor. *Revista Argentina de Ciencias del Comportamiento* 3: 1–11.

Aitchison, Jean. 1996. *The seeds of speech: Language origin and evolution.* Cambridge, U.K.: Cambridge University Press.

Aksu-Koc, Ayjan A., and Dan Slobin. 1985. The acquisition of Turkish. In Dan Slobin, ed., *The crosslinguistic study of acquisition data:* vol. 1, *The data,* 525–593. Hillsdale, N.J.: Lawrence Erlbaum.

Alexander, Garrett E., and Michael D. Crutcher. 1990. Functional architecture of basal ganglia circuits: Neural substrates of parallel processing. *Trends in Neurosciences,* 13: 266–271.

References

Alleyne, Mervyn. 1971. Acculturation and the cultural matrix of creolization. In Dell Hymes, ed., *Pidginization and creolization of languages,* 169–186. Cambridge, U.K.: Cambridge University Press.

Allott, R. 1992. The motor theory of language: Origin and function. In J. Wind, B. H. Bichakjian, A. Nocentini, and B. Chiarelli, eds., *Language origin: A multidisciplinary approach,* 105–119. Dordrecht: Kluwer.

Allport, D. A. 1985. Distributed memory, modular subsystems and dysphasia. In S. K. Newman and R. Epstein, eds., *Current perspectives in dysphasia,* 32–60. Edinburgh: Churchill Livingston.

Ambridge, B., C. Rowland, A. Theakston, and M. Tomasello. 2006. Comparing different accounts of inversion errors in children's non-subject wh-questions: What experimental data can tell us. *Journal of Child Language* 33: 519–557.

Ambrose, S. H. 2001. Paleolithic technology and human evolution. *Science* 291: 1748–1753.

Anisfeld, Moshe, Erica S. Rosenberg, Mara J. Hoberman, and Don Gasparini. 1998. Lexical acceleration coincides with the onset of combinatorial speech. *First Language* 18: 165–184.

Ansaldo, Umberto, and Stephen Matthews. 2007. Deconstructing creole: The rationale. In Umberto Ansaldo, Stephen Matthews, and Lisa Lim, eds., *Deconstructing creole,* 1–18. Amsterdam: Benjamins.

Arabidopsis Genome Initiative. 2000. Analysis of the genome sequence of the flowering plant Arabidopsis thaliana. *Nature* 408: 796–815.

Arbib, Michael. 2003. Schema theory. In M. A. Arbib, ed., *The handbook of brain theory and neural networks,* 2nd edition, 993–998. Cambridge, Mass.: MIT Press.

———. 2005. From monkey-like action recognition to human language. *Behavioral and Brain Sciences* 28: 105–167.

Arbib, Michael, and Derek Bickerton, eds. 2010. *The emergence of protolanguage: Holophrasis vs. compositionality.* Amsterdam: Benjamins.

Archetti, M., and I. Scheuring. 2012. Review: Game theory of public goods in one-shot social dilemmas without assortment. *Journal of Theoretical Biology* 299: 9–20.

Arends, Jacques. 1989. Syntactic developments in Sranan: Creolization as a gradual process. Ph.D. diss., Catholic University of Nijmegen.

———. 1995. Demographic factors in the formation of Sranan, In J. Arends, ed., *The early stages of creolization,* 233–277. Amsterdam: Benjamins.

Arnold, Kate, and Klaus Zuberbuhler. 2006. The alarm-calling system of adult male putty-nosed monkeys, *Cercopithecus nictitans martini. Animal Behavior* 72: 643–653.

Assadolahi, Ramin, and Friedeman Pulvermuller. 2001. Neural network classification of word-evoked neuromagnetic brain activity. In Stefan Wermter, Jim Austin, and David Willshaw, eds., *Emergent neural computational architectures based on neuroscience,* 311–319. Berlin: Springer-Verlag.

Aureli, F., C. M. Schaffner, C. Boesch, S. K. Bearder, J. Call, C. A. Chapman, R. Connor, A. Di Fiore, R. I. M. Dunbar, S. P. Henzi, K. Holekamp, A. Korstjen, R. H. Layton, P. Lee, J. Lehmann, J. H. Manson, G. Ramos-Fernandez, and C. P. Van Schaik. 2008. Fission-fusion dynamics: New research frameworks. *Current Anthropology* 49: 627–665.

Austin, Peter A. 2001. Word order in a free word order language: The case of Jiwarli. In Jane Simpson, David Nash, Mary Laughren, Peter Austin, and Barry Alpher, eds., *Forty years on: Ken Hale and Australian languages*, 205–323. Canberra: Pacific Linguistics.

Austin, Peter A., and Joan Bresnan. 1996. Non-configurationality in Australian Aboriginal languages. *Natural Language and Linguistic Theory* 14: 215–268.

Avital, E., and E. Jablonka. 2000. *Animal traditions: Behavioural inheritance in evolution.* Cambridge, U.K.: Cambridge University Press.

Axelrod, R. 1984. *The evolution of cooperation.* New York: Basic Books.

Ayala, Francisco J., and Theodosius Dobzhansky, eds. 1974. *Studies in the philosophy of biology: Reduction and related problems.* Berkeley: University of California Press.

Aydede, Murat. 2010. The language of thought hypothesis. In Edward N. Zalta, ed., *The Stanford encyclopedia of philosophy.* Stanford: Stanford University.

Ayoun, Dalila. 2003. *Parameter setting in language acquisition.* London: Continuum.

Bailey, K. G. D., and F. Ferreira. 2003. Disfluencies affect the parsing of garden-path sentences. *Journal of Memory and Language* 4: 183–200.

Bailey, V. 1923. The combing claws of the beaver. *Journal of Mammalogy* 4: 77–79.

Baker, C. L. 1979. Syntactic theory and the projection principle. *Linguistic Inquiry* 10: 533–561.

Baker, M. C. 1989. Object sharing and projection in serial verb constructions. *Linguistic Inquiry* 20: 513–553.

———. 2001. *The atoms of language.* New York: Basic Books.

———. 2002. On zero agreement and polysynthesis. Unpublished manuscript, Rutgers University.

———. 2003. Linguistic differences and language design. *Trends in Cognitive Sciences* 7: 349–353.

Baker, Philip. 1995. Some developmental inferences from historical studies of pidgins and creoles. In J. Arends, ed., *The early stages of creolization*, 1–24. Amsterdam: Benjamins.

Baker, Philip, and Chris Corne. 1982. *Isle de France Creole: Affinities and origins.* Ann Arbor, Mich.: Karoma.

Bakker, Peter. 1989. "The language of the coast tribes is half Basque": A Basque-Amerindian pidgin in use between Europeans and Native Americans in North America, ca. 1540–ca. 1640. *Anthropological Linguistics* 31: 117–147.

References

Bakker, Peter, Norval Smith, and Tonjes Veenstra. 1995. Saramaccan. In J. Arends, P. Muuysken, and N. Smith, eds., *Pidgins and Creoles: An introduction*, 165–178. Amsterdam: Benjamins.

Balari, Sergio, and Guillermo Lorenzo. 2010. Specters of Marx: A review of *Adam's tongue* by Derek Bickerton. *Biolinguistics* 4: 116–127.

Baldwin, J. M. 1896. A new factor in evolution. *American Naturalist* 30: 441–451, 533–536.

Ballard, Dana H. 1986. Cortical connections and parallel processing: Structure and function. *Behavioral and Brain Sciences* 9: 67–90.

Baptiste, Marlyse. 2012. On universal grammar, the bioprogram hypothesis, and creole genesis. *Journal of Pidgin and Creole Studies* 27: 351–376.

Barkow, J., L. Cosmides, and J. Tooby, eds. 1992. *The adapted mind: Evolutionary psychology and the generation of culture*. New York: Oxford University Press.

Barrington, D. 1773. Experiments and observations on the singing of birds. *Philosophical Transactions of the Royal Society of London* 63: 249–291.

Barss, Andrew, and Howard Lasnik. 1986. A note on anaphora and double objects. *Linguistic Inquiry* 17: 347–354.

Battye, Adrian, and Ian Roberts. 1995. *Clause structure and language change*. Oxford: Oxford University Press.

Belletti, Adriana, and Luigi Rizzi. 2000. An interview on minimalism, with Noam Chomsky. University of Siena, November 8–9, 1999; revised March 16, 2000. http://www.cse.iitk.ac.in/users/hk/cs789/papers/Belletti-Rizzi-interview-Chomsky02.pdf/

Bellugi, Ursula. 1967. The acquisition of negation. Ph.D. diss., Harvard University.

Beltman, J. B., P. Haccou, and C. ten Cate. 2004. Learning and colonization of new niches: A first step towards speciation. *Evolution* 58: 35–46.

Bennett, Maxwell, and Peter Hacker. 2007. The conceptual presuppositions of cognitive neuroscience. In M, Bennett, D, Dennett, P. Hacker, and J. Searle, eds., *Neuroscience and philosophy: Brain, mind and language*, 127–138. New York: Columbia University Press.

Berwick, Robert C., and Noam Chomsky. 2011. The biolinguistic program: The current stage of its development. In Anna Maria di Sciullo and Cedric Boeckx, eds., *The biolinguistic enterprise*, 19–41. Oxford: Oxford University Press.

Berwick, Robert C., and A. S. Weinberg. 1984. *The grammatical basis of linguistic performance: Language use and acquisition*. Cambridge, Mass.: MIT Press.

Bever, Thomas G. 2009. Biolinguistics today and Platonism yesterday. In W D. Lewis, S. Karimi, H. Harley, and S. O. Farrar, eds., *Time and again*, 227–232. Amsterdam: Benjamins.

Bickerton, Derek. 1971. Inherent variability and variable rules. *Language* 47: 457–492.

———. 1973. On the nature of a creole continuum. *Language* 49: 640–699.

———. 1974. Creolization, linguistic universals, natural semantax and the brain. *Working Papers in Linguistics* (University of Hawaii) 6: 125–141.

———. 1975/2009. *Dynamics of a Creole system.* Cambridge, U.K.: Cambridge University Press.

———. 1977. *Creole syntax.* Mimeo, University of Hawaii.

———. 1981. *Roots of language.* Ann Arbor, Mich.: Karoma.

———. 1984a. Creole still is king: Response to commentary. *Behavioral and Brain Sciences* 7: 212–221.

———. 1984b. The language bioprogram hypothesis. *Behavioral and Brain Sciences* 7: 173–188.

———. 1986. Beyond *Roots*: progress or regress? *Journal of Pidgin and Creole Languages* 1: 135–140.

———. 1987. Creoles and West African languages: A case of mistaken identity? In P. Muysken and N. V. Smith, eds., *Universals versus substrata in creole genesis,* 25–40. Amsterdam: Benjamins,

———. 1988. Creole languages and the bioprogram. In F. J. Newmeyer, ed., *Linguistics: The Cambridge survey,* vol. 2, 268–284. Cambridge, U.K.: Cambridge University Press.

———. 1989. Seselwa serialization and its significance. *Journal of Pidgin and Creole Languages* 4: 155–183.

———. 1990a. Haitian demographics and creole genesis. *Canadian Journal of Linguistics* 35: 217–219.

———. 1990b. *Language and species.* Chicago: University of Chicago Press.

———. 1991. On the supposed "gradualness" of creole development. *Journal of Pidgin and Creole Languages* 6: 25–58.

———. 1992a. The sociohistorical matrix of creolization. *Journal of Pidgin and Creole Languages* 7: 307–18

———. 1992b. Unified cognitive theory: You can't get there from here. *Bejavioral and Brain Sciences* 437–438.

———. 1994. The origins of Saramaccan syntax: A reply to McWhorter. *Journal of Pidgin and Creole Languages* 9: 65–78.

———. 1995. *Language and human behavior.* Seattle: University of Washington Press.

———. 1996. Why serial verb constructions in "Isle-de-France Creole" can have subjects: A reply to Corne, Coleman and Curnow. In P. Baker, ed., *Changing meanings, changing functions,* 155–169. London: University of Westminster Press.

———. 1999. How to acquire language without positive evidence: What acquisition can learn from creoles. In M. DeGraff, ed., *Language creation and language change,* 49–74. Cambridge, Mass.: MIT Press.

———. 2000. How protolanguage became language. In C. Knight, M. Studdert-Kennedy, and J. R. Hurford, eds., *The evolutionary emergence of language,* 264–284. Cambridge, U.K.: Cambridge University Press.

———. 2006. Creoles, capitalism, and colonialism. In J. Clancy Clements, T. A. Klingler, D. Piston-Hatlen, and K. J. Rottet, eds., *History, society and variation*, 137–152. Amsterdam: John Benjamins.
———. 2008. *Bastard tongues*. New York: Hill & Wang.
———. 2009. *Adam's tongue*. New York: Hill & Wang.
Bickerton, Derek, and Carol Odo. 1976. *General phonology and pidgin syntax*. Mimeo, University of Hawaii.
Bickerton, Derek, and Eors Szathmary. 2011. Confrontational scavenging as a possible source for language and cooperation. *BMC Evolutionary Biology* 11: 261.
Binford, L. S. 1985. Human ancestors: Changing views of their behavior. *Journal of Anthropological Archaeology* 4: 292–327.
Bloom, Lois. 1970. *Language development: Form and function in emerging grammars*. Cambridge, Mass.: MIT Press.
Blumenschine, R. J. 1987. Characteristics of an early hominid scavenging niche. *Current Anthropology* 28: 383–407.
Blumenschine, R. J., J. V. Cavallo, and S. P. Capaldo. 1994. Competition for carcasses and early hominid behavioral ecology. *Journal of Human Evolution* 27: 197–214.
Bobe, R., A. K. Behrensmeyer, and R. E. Chapman. 2002. Faunal change, environmental variability and late Pliocene hominin evolution. *Journal of Human Evolution*, 42: 475–497
Bock, J. Kathryn. 1986. Meaning, sound, and syntax: Lexical priming in sentence production. *Journal of Experimental Psychology: Learning, Memory, and Cognition* 12: 575–586.
Boeckx, Cedric. 2008. Islands. *Language and Linguistics Compass* 2: 151–167.
———. 2010. What principles and parameters got wrong. Unpublished manuscript, ICREA/UAB.
———. 2012. The emergence of language from a biolinguistic point of view. In M. Tallerman and K. R. Gibson, eds., *The Oxford handbook of language evolution*, 492–501. Oxford: Oxford University Press.
———, forthcoming, *Elementary syntactic structures*. Cambridge. U.K.: Cambridge University Press.
Boesch, Christopher. 1994. Hunting strategies of Gombe and Thai chimpanzees. In R. W. Wrangham, ed., *Chimpanzee culture*. Cambridge, Mass.: Harvard University Press.
Boesch, Christopher, and Hedwige Boesch. 1989. Tool use and tool making in wild chimpanzees. *Folia Primatologia* 54: 86–99.
Bollee, Annegret. 1977. *Le creole francaise des Seychelles*. Tubingen: Niemeyer.
Bonner, J. T. 1983. *The evolution of culture in animals*. Princeton, N.J.: Princeton University Press.
Borer, Hagit. 1984. *Parametric syntax: Case studies in Semitic and Romance languages*. Dordrecht: Foris.

Borer, Hagit, and Ken Wexler. 1987. The maturation of syntax. In T. Roeper and E. Williams, eds., *Parameter setting,* 123–172. Dordrecht: Reidel.

———. 1992. Bi-unique relations and the maturation of grammatical principles. *Natural Language and Linguistic Theory* 10: 147–187.

Borges, Jorge-Luis. 1975. *A universal history of infamy.* Trans. Norman Thomas de Giovanni. London: Penguin Books.

Bornstein, M. H., W. Kessen, and S. Weiskopf. 1976. Color vision and hue categorization in young human infants. *Journal of Experimental Psychology* 2: 115–129.

Boroditsky, Lera, and Jesse Prinz. 2008. What thoughts are made of. In Gun T. Semin and Eliot R. Smith, eds., *Embodied grounding: Social, cognitive, affective and neuroscientific approaches,* 98–110. Cambridge, U.K.: Cambridge University Press.

Boser, K., B. Lust, L. Santelmann, and J. Whitman. 1992. The syntax of CP and V-2 in early child German: The strong continuity hypothesis. In *Proceedings of the North East Linguistic Society* 22: 51–65.

Botha, Rudolf. 2012. Protolanguage and the "God particle." *Lingua* 122: 1308–1324.

Bowerman, Melissa. 1974. Learning the structure of causative verbs: A study in the relationship of cognitive, semantic and syntactic development. *Papers and Reports on Child Language Development* (Stanford University) 8: 142–178.

———. 1982. Reorganizational processes in lexical and syntactic development. In L. Gleitman and E. Wanner, eds., *Language acquisition: The state of the art,* 319–348. Cambridge, U.K.: Cambridge University Press.

———. 1988. The "no negative evidence" problem: How do children avoid constructing an overgeneral grammar? In J. A. Hawkins, ed., *Explaining language universals,* 73–101. Oxford: Blackwell.

Boyd, Robert, Herbert Bowles, Samuel Gintis, and Peter J. Richardson. 2003. The evolution of altruistic punishment. *Proceedings of the National Academy of Science* 100: 3531–3535.

Brain, Robert R. 1961. The neurology of language. *Proceedings of the Royal Society of Medicine* 54: 433–441.

Braine, Martin D. S. 1963. The ontogeny of English phrase structure: The first phase. *Language* 39: 1–13.

———. 1976. Children's first word combinations. *Monographs of the Society for Research in Child Development* 41, no. 1 (entire issue).

Bramble, Dennis M., and Daniel E. Lieberman. 2004. Endurance running and the evolution of *Homo. Nature* 432: 345–352.

Branigan, G. 1979. Some reasons why successive word utterances are not. *Journal of Child Language* 6: 411–421.

Braun, Maria. 2009. *Word formation and creolisation: The case of Early Sranan.* Tubingen: Niemeyer.

Breivik, Leiv Egil, and Ernst Hakon Jahr, eds. 1989. *Language change: Contributions to the study of its causes.* Berlin: Mouton de Gruyter.

Brenner, Sydney. 1974. The genetics of *Caenorhabditis Elegans. Genetics* 77: 71–94.

Brereton, Bridget. 2005. Family strategies, gender and the shift to wage labor in the British Caribbean. In Pamela Scully and Diana Paton Duke, eds., *Gender and slave emancipation in the Atlantic world,* 143–161. Durham, N.C.: Duke University Press.

Broch, Ingvild, and Ernst Jahr. 1984. Russenorsk: A new look at the Russo-Norwegian pidgin in northern Norway. In P. S. Ureland and I. Clarkson, eds., *Scandinavian Language Contacts,* 21–65. Cambridge, U.K.: Cambridge University Press.

Brody, Michael. 2002. On the status of derivations and representations. In S. Epstein, ed., *Derivation and Explanation in the Minimalist Program,* 19–42. Oxford: Blackwell.

Brown, Roger. 1973. *A first language: The early stages.* Cambridge, Mass.: Harvard University Press.

Brown, Roger, and Colin Fraser. 1964. The acquisition of syntax. In U. Bellugi and R. Brown, eds., *The acquisition of language,* 43–79. Chicago: University of Chicago Press.

Brown, Roger, and C. Hanlon. 1970. Derivational complexity and order of acquisition in child speech. In Roger Brown, ed., *Psycholinguistics,* 11–54. New York: Free Press.

Buchler, J. 1955. *The philosophical writings of Peirce.* New York: Dover.

Bunn, H. T., and E. M. Kroll. 1986. Systematic butchery by Plio-Pleistocene hominds at Olduvai Gorge, Tanzania. *Current Anthropology* 27: 431–452.

Burling, Robbins. 2005. *The talking ape: How language evolved.* Oxford: Oxford University Press.

Byrne, R. W. 2007. Culture in great apes: Using intricate complexity in feeding skills to trace the evolutionary origin of human technical prowess. *Philosophical Transactions of the Royal Society B* 362: 577–585.

Calvin, William H. 1982. Did throwing stones shape hominid brain evolution? *Ethology and Sociobiology* 3: 115–124.

———. 1996. *The cerebral code: Thinking a thought in the mosaics of the mind.* Cambridge, Mass.: MIT Press.

Calvin, William H., and Derek Bickerton. 2000. *Lingua ex machina.* Cambridge, Mass.: MIT Press.

Cangelosi, Angelo. 2001. Evolution of communication and language using signals, symbols, and words. *Transactions on Evolutionary Computation* 5: 93–101.

Caramazza, A., and A. E. Hillis. 1991. Lexical organization of nouns and verbs in the brain. *Nature* 349: 788–790.

Caramazza, A., A. E. Hillis, B. C. Rapp, and C. Romani. 1990. The multiple semantic hypothesis: Multiple confusions? *Cognitive Neuropsychology* 7: 161–189.

Carey, S., and E. Bartlett. 1978. Acquiring a single new word. *Proceedings of the Stanford Child Language Conference* 15: 17–29.

Carlson, Mary, and W. I. Welker. 1976. Some morphological, physiological and behavioral specializations in North American beavers *(Castor canadensis)*. *Brain, Behavior and Evolution* 13: 302–326.

Carr, Elizabeth B. 1972. *Da kine talk: From pidgin to Standard English in Hawaii.* Honolulu: University of Hawaii Press.

Carranza, J. A., A. Escudero, and A. G. Brito 1991. De las palabras aisladas a las combinaciones de palabras. *Anales de psicología* 72: 163–180.

Carroll, Sean B. 2005. *Endless forms most beautiful: The new science of evo devo.* New York: Norton.

———. 2008. Evo-Devo and an expanding evolutionary synthesis: A genetic theory of morphological evolution. *Cell 134*, 25–36.

Carstairs-McCarthy, Andrew. 1999. *The origins of complex language.* Oxford: Oxford University Press.

———. 2011. The evolutionary relevance of more and less complex forms of language. In M. Tallerman and K. R. Gibson, eds., *The Oxford handbook of language evolution,* 435–441. Oxford: Oxford University Press.

Catani, Marco, and Marsel Mesulam, 2008. The arcuate fasciculus and the disconnection theme in language and aphasia: History and current state. *Cortex* 44 (8): 953–961.

C. elegans Sequencing Consortium. 1998. Genome sequence of the nematode *C. elegans:* A platform for investigating biology. *Science* 282: 2012–2018.

Chapman, C. A., R. W. Wrangham, and L. J. Chapman. 1995. Ecological constraints on group size: An analysis of spider monkey and chimpanzee subgroups. *Behavioral Ecology and Sociobiology* 36: 59–70.

Chater, Nick, and Cecilia Hayes. 1994. Animal concepts: Content and discontents. *Mind and Language* 9: 209–246.

Chaudenson, Robert. 1981. *Textes anciens en créole réunionnais et mauricien: Comparaison et essai d'analyse.* Hamburg: H. Buske.

———. 1989. Linguistic creolization, cultural creolization. *Etudes Créoles* 12: 53–57.

———. 1992. *Des îsles, des hommes, des langues.* Paris: L'Harmattan.

———. 2001. *Creolization of language and culture.* (Revised version of Chaudenson 1992, in collaboration with Salikoko Mufwene). London: Routledge.

Cheney, D. L., and R. M. Seyfarth. 1988. Assessment of meaning and the detection of unreliable signals by vervet monkeys. *Animal Behavior* 36: 477–486.

———. 1990. *How monkeys see the world.* Chicago: University of Chicago Press.

———. 2010. Primate communication and human language: Continuities and discontinuities. In P. M. Kappeler and J. B. Silk, eds., *Mind the gap,* 283–298. Berlin: Springer Verlag.

Cherniak, Christopher. 1994. Philosophy and computational neuroanatomy. *Philosophical Studies* 73: 89–107.

———. 2005. Innateness and brain-wiring optimization: Non-genomic nativism. In Antonio Zilhao, ed., *Evolution, rationality and cognition*, 103–112. New York: Routledge.

Cherniak, Christopher, Z. Mokhtarzada, and Uri Nodelman. 2002. Optimal wiring models of neuroanatomy. In Giorgio A. Ascoli, ed., *Computational neuroanatomy: Principles and methods*, vol. 1, 71–82. New York: Humana Press.

Chittka, Lars, and Keith Jensen. 2011. Animal cognition: Concepts from apes to bees. *Current Biology* 21: R116–R119.

Chomsky, Noam. 1957. *Syntactic structures*. The Hague: Mouton.

———. 1959. Review of Skinner, *Verbal behavior*. *Language* 35: 26–57.

———. 1965. *Aspects of the theory of syntax*. Cambridge, Mass.: MIT Press.

———. 1966. *Cartesian linguistics*. New York: Harper & Row.

———. 1968. *Language and mind*. New York: Harcourt, Brace & World.

———. 1972. *Language and mind*. Enlarged edition. New York: Harcourt, Brace, Jovanovich.

———. 1980. *Rules and representations*. New York: Columbia University Press.

———. 1981. *Lectures on government and binding: The Pisa lectures*. Dordrecht: Foris.

———. 1983. Things no amount of learning can teach. Interview with John Gliedman, *Omni* 6:11, November.

———. 1986. *Knowledge of language: Its nature, origin, and use*. New York: Praeger.

———. 1988. *Language and problems of knowledge: The Managua lectures*. Cambridge, Mass.: MIT Press.

———. 1991. Some notes on economy of derivation and representation. In Robert Freidin, ed., *Principles and parameters in comparative grammar*, 417–454. Cambridge, Mass.: MIT Press.

———. 1994. Bare phrase structure. *MIT Occasional Papers in Linguistics* No. 5.

———. 1995. *The minimalist program*. Cambridge, Mass.: MIT Press.

———. 1999. Derivation by phase. *MIT Occasional Papers in Linguistics* No. 18.

———. 2000. Minimalist inquiries: The framework. In Robert Martin, David Michaels, and Juan Uriagereka, eds., *Step by step: Essays on minimalist syntax in honor of Howard Lasnik*, 89–155. Cambridge, Mass.: MIT Press.

———. 2002. *On nature and language*. Cambridge, U.K.: Cambridge University Press.

———. 2005. Three factors in language design. *Linguistic Inquiry* 36: 1–22.

———. 2007. Of minds and language. *Biolinguistics* 1: 9–27.

———. 2010. Some simple evo-devo theses: How true might they be for language? In R. K. Larson, V. Deprez, and H. Yamakido, eds., *The evolution of human language: Biolinguistic perspectives*, 45–62. Cambridge, U.K.: Cambridge University Press.

Christiansen, M. H., and N. Chater. 2008. Language as shaped by the brain. *Behavioral and Brain Sciences* 31: 489–509.

Christiansen, M. H., and Simon Kirby. 2003. Introduction to M. H. Christiansen and S. Kirby, eds., *Language evolution.*, 1–15. Oxford: Oxford University Press.

Christianson, K., A. Hollingworth, J. F. Halliwell, and F. Ferreira. 2001. Thematic roles assigned along the garden path linger. *Cognitive Psychology* 42: 368–407.

Clayton, Nicola S., and Anthony Dickinson. 1998. Episodic-like memory during cache recovery by scrub jays. *Nature* 395: 272–274.

Clayton, Nicola S, D. P. Griffiths, N. J. Emery, and A. Dickinson. 2001. Elements of episodic–like memory in animals. *Philosophical Transactions of the Royal Society of London. B 29;* 356 (1413): 1483–1491.

Clutton-Brock, T. H., and G. A. Parker. 1995. Punishment in animal societies. *Nature* 373: 209–216.

Coelho, A. 1880–86. Os dialectos romanicos ou neo-latinos na Africa, Asia e America. *Boletin do Sociedade de Geographia de Lisboa* 2: 129–196, 3: 451–478, 6: 705–755.

Collins, J. 2010. Complexity and simplicity: Continuity and discontinuity in generative linguistics. http://eastanglia.academia.edu/JohnCollins/Papers/91510/.

Conway, M. A. 2005. Memory and the self. *Journal of Memory and Language* 53: 594–628.

Conway, M. A., and C. W. Pleydell-Pearce. 2000. The construction of autobiographical memories in the self-memory system. *Psychological Review* 107: 261–288.

Conway Morris, Simon. 2003. *Life's solution: Inevitable humans in a lonely universe.* Cambridge, U.K.: Cambridge University Press.

———. 2008. *The deep structure of biology.* West Conshohocken, Penn.: Templeton Press.

Corne, Chris. 1977. *Seychelles Creole grammar.* Tubingen: Narr.

———. 1988. Mauritian Creole reflexives. *Journal of Pidgin and Creole Languages* 3: 69–94.

———. 1995. Nana k nana, nana k napa: The paratactic and hypotactic relative clauses of Reunion Creole. *Journal of Pidgin and Creole Languages* 10: 57–76.

Cowen, Ron. 2013. Small stars host droves of life-friendly worlds. *Nature/News,* February 9.

Crain, Stephen. 1991. Language acquisition in the absence of experience. *Behavioral and Brain Sciences* 14: 597–612.

Crain, Stephen, and Diane Lillo-Martin. 1999. *An introduction to linguistic theory and language acquisition.* Oxford: Blackwell.

Cross, T. G. 1977. Mothers' speech adjustments: The contribution of selected child listener variables. In C.E. Snow and C.A. Ferguson, eds., *Talking to*

children: Language input and acquisition, 151–188. Cambridge, U.K.: Cambridge University Press.
Curtiss, Susan. 1977. *Genie: A Psycholinguistic Study of a Modern-Day "Wild Child."* Boston, Mass.: Academic Press
Daeleman, Jan. 1972. Kongo elements in Saramaccan Tongo. *Journal of African Languages* 11: 1–44.
Dally, Joanna M., Nathan J. Emery, and Nicola S. Clayton. 2006. Food-caching western scrub-jays keep track of who was watching when. *Science* 16: 1662–1665.
Damasio, A. R. 1989. The brain binds entities and events by multiregional activation from convergence zones. *Neural Computation* 1: 123–132.
Damasio, A. R., and D. Tranel. 1993. Nouns and verbs are retrieved with differentially distributed neural systems. *Proceedings of the National Academy of Science* 90: 4957–4960.
Damasio, H., T. J. Grabowski, D. Tranel, R. D. Hichwa, and A. R. Damasio. 1996. A neural basis for lexical retrieval. *Nature* 380: 499–505.
Darwin, Charles. 1871. *The descent of man, and selection in relation to sex.* London: John Murray.
Dawkins. Richard. 1976. *The selfish gene.* Oxford: Oxford University Press.
———. 1986. *The blind watchmaker.* New York: Norton.
Deacon, Terrence. 1997. *The symbolic species.* New York: Norton.
———. 2006. The evolution of language systems in the human brain. In John Kaas, ed., *Evolution of nervous systems:* vol. 5, *The evolution of primate nervous systems,* 897–916. Amsterdam: Elsevier.
———. 2010. A role for relaxed selection in the evolution of the language capacity. *Proceedings of the National Academy of Science,* Supplement 2: 9000–9006.
DeCamp, David. 1971. Towards a generative analysis of a post-creole speech continuum. In Dell Hymes, ed., *Pidginization and creolization of languages,* 349–370. Cambridge, U.K.: Cambridge University Press.
Decety, J., M. Jeannerod, M. Germain, J. Pastene. 1991. Vegetative response during imagined movement is proportional to mental effort. *Behavior and Brain Research* 42: 1–5.
Decety, J., D. Perani, M. Jeannerod, V. Bettinardi, B. Tadary, R. Woods, J. C. Mazziotta, and F. Fazio. 1994. Mapping motor representations with positron emission tomography. *Nature* 371: 600–602.
DeGraff, Michel, ed. 1999. *Language creation and language change: Creolization, diachrony and development.* Cambridge, Mass.: MIT Press.
———. 2003. Against creole exceptionalism. *Language* 79: 391–410.
———. 2005. Linguists' most dangerous myth: The fallacy of Creole exceptionalism. *Language in Society* 34: 533–591.
———. 2007. Haitian Creole. In John Holm and Peter Patrick, eds., *Comparative creole syntax: Parallel outlines of 18 creole grammars,* 101–126. London: Battlebridge.

Dejean, A., P. J. Solano, J. Ayroles, B. Corbara, and J. Orivel. 2005. Arboreal ants build traps to capture prey. *Nature* 434: 973.

De la Torre Sainz, Ignacio, and Manuel Dominguez-Rodrigo. 1998. Gradualismo y equilibrio puntuado en el origen del comportamiento humano. *Zephyrus* 51: 13–18.

Deprez, V., and A. Pierce. 1993. Negation and functional projections in early grammar. *Linguistic Inquiry* 24: 25–67.

Dere E., E. Kart-Teke, J. P. Huston, and M. A. De Souza Silva. 2006. The case for episodic memory in animals. *Neuroscience Biobehavioral Review* 30: 1206–1224.

Dessalles, J. L. 2000. Language and hominid politics. In C. Knight, M. Studdert-Kennedy, and J. Hurford, eds., *The evolutionary emergence of language,* 62–80. Cambridge, U.K.: Cambridge University Press.

——. 2009. *Why we talk: The evolutionary origins of language.* 2nd edition. Oxford: Oxford University Press.

——. 2010. *Providing information can be a stable non-cooperative evolutionary strategy.* Paris: Technical Report, Telecom Paris Tech 2010D025.

Dieckmann, U., and M. Doebeli. 1999. On the origin of species by sympatric speciation. *Nature* 400: 354–357.

Diller, K. C., and R. L. Cann. 2010. The innateness of language: A view from genetics. In Andrew D. M. Smith, Marieke Schouwstra, Bart de Boer, and Kenny Smith, eds., *Proceedings of the 8th International Conference on the Evolution of Language,* 107–115. Singapore: World Scientific.

Dimmendaal, Gerrit J. 2001. Areal diffusion versus genetic inheritance: An African perspective. In Alexandra Aikhenvald and R. M. W. Dixon, eds., *Areal diffusion and genetic inheritance: Problems in comparative linguistics,* 358–392. Oxford: Oxford University Press.

Dirven, Renee, and Marjojlin Verspoor. 2004. *Cognitive exploration of language and linguistics.* Amsterdam: Benjamins.

Dogil, G., H. Ackermann, W. Grodd, H. Haider, H. Kamp, J. Mayer, A. Riecker, and D. Wildgruber. 2002. The speaking brain: A tutorial introduction to fMRI experiments in the production of speech, prosody and syntax. *Journal of Neurolinguistics* 15: 59–90.

Dominguez-Rodrigo, M., T. R. Pickering, S. Semaw, and M. J. Rogers. 2005. Cutmarked bones from Pliocene archaeological sites at Gona, Afar, Ethiopia: Implications for the function of the world's oldest stone tools. *Journal of Human Evolution* 48: 109–121.

Dopke, S. 1998. Competing language structures: The acquisition of verb placement by bilingual German-English children. *Journal of Child Language* 25: 555–584.

Dornhaus, A., and L. Chittka. 1999. Insect behavior: Evolutionary origins of bee dances. *Nature* 401: 38.

Dresher, Bezalel Elan. 1999. Charting the learning path: Cues to parameter setting. *Linguistic Inquiry* 30: 27–67.

Dryer, Matthew S. 2002. Case distinctions, rich verb agreement, and word order type. *Theoretical Linguistics* 28: 151–157.

Dunbar, Robin J. M. 1996. *Grooming, gossip and the evolution of language.* London: Faber and Faber.

Dunn, M., S. Greenhill, S. Levinson, and R. Gray. 2011. Evolved structure of language shows lineage-specific trends in word-order universals. *Nature* 473: 79–82.

Dyer, Fred C. 2002. The biology of the dance language. *Annual Review of Entomology* 47: 917–949.

Dyer, Fred C., and J. M. Gould. 1983. Honey-bee navigation. *American Scientist* 71: 587–597.

Dyson, Freeman. 1999. *Origins of life.* Cambridge, U.K.: Cambridge University Press.

Eacott, M. J., A. Easton, and A. Zinkivskay. 2005. Recollection in an episodic-like memory task in the rat. *Learning and Memory* 12: 221–223.

Eaton, Stanley B. III, and Loren Cordain. 2002. Evolution, diet and health. In Peter S. Ungar and Mark Franklyn Teaford, eds., *Human diet: Its origin and evolution,* 7–18. Westport, Conn.: Greenwood.

Eberhard, W. G. 1990. Function and phylogeny of spider webs. *Annual Review of Ecology and Systematics* 21: 341–372.

Edelman, Gerald. 1987. *Neural Darwinism: The theory of neuronal group selection.* New York: Basic Books.

Eldredge, Niles, and Stephen Jay Gould. 1972. Punctuated equilibrium: An alternative to phyletic gradualism. In T. J. M. Schopf, ed., *Models in paleobiology,* 82–115. San Francisco: Freeman Cooper.

Emery, N. J., and N. S. Clayton. 2001. Effects of experience and social context on prospective caching strategies by scrub jays. *Nature* 414: 443–446.

Emonds, Joseph. 1985. *A unified theory of syntactic categories.* Dordrecht: Foris.

Epstein, Samuel D. 1999. Unprincipled syntax: The derivation of syntactic relations. In S. D. Epstein and N. Hornstein, eds., *Working minimalism,* 317–345. Cambridge, Mass.: MIT Press.

Epstein, Samuel D., Eric M. Groat, Ruriko Kawashima, and Hisatsugu Kitahara. 1998. *A derivational approach to syntactic relations.* New York: Oxford University Press.

Evans, C. S., L. Evans, and P. Marler. 1993. On the meaning of alarm calls: Functional reference in an avian vocal system. *Animal Behaviour* 46: 23–38.

Evans, Nicholas, and Stephen C. Levinson. 2009a. The myth of language universals: Language diversity and its importance for cognitive science. *Behavioral and Brain Sciences* 32: 429–448.

———. 2009b. With diversity in mind: Freeing the language sciences from universal grammar. *Behavioral and Brain Sciences* 32: 472–492.

Eyal, S. 1976. Acquisition of question-utterances by Hebrew-speaking children. MA thesis, Tel Aviv University.

Falk, Dean. 1990. Brain evolution in Homo: The "radiator" theory. *Behavioral and Brain Sciences* 13: 333–344.

Fauconnier, Giles, and Mark Turner. 2002. *The way we think: Conceptual blending and the mind's hidden complexities.* New York: Basic Books.

Fedor, A., P. Ittzés, and E. Szathmáry. 2010. Parsing recursive sentences with a connectionist model including a neural stack and synaptic gating. *Journal of Theoretical Biology* 271: 100–105.

Fey, M. E., and D. F. Loeb. 2002. An evaluation of the facilitative effects of inverted yes-no questions on the acquisition of auxiliary verbs, *Journal of Speech, Language and Hearing Research* 45: 160–174.

Finlay, Barnara L., and Dale R. Sengelaub. 1981. Toward a neuroethology of mammalian vision: Ecology and anatomy of rodent visuomotor behavior. *Behavioural and Brain Research* 3: 133–149.

Fishbein, William. 2000. The case against memory consolidation in REM sleep: Balderdash! *Behavioral and Brain Sciences* 23: 934–936.

Fitch, W. Tecumseh. 2000. The evolution of speech: A comparative review. *Trends in Cognitive Sciences* 4: 258–266.

———. 2010. *The evolution of language.* Cambridge, U.K.: Cambridge University Press.

———. 2011. An instinct to learn. http://www.edge.org/response-detail/11453

Fleck, D. 2007. Evidentiality and tense in Matses. *Language* 8: 589–614.

Fodor, J. A., and M. Garrett. 1967. Some syntactic determinants of sentential complexity. *Perception and Psychophysics* 2: 289–296.

Foley, Robert. 1987. *Another unique species: Patterns in human evolutionary ecology.* Harlow, U.K.: Longman.

Foster, K. R., and F. L. W. Ratnieks. 2005. A new eusocial vertebrate? *Trends in Ecology and Evolution* 20: 363–364.

Francis, W. Nelson, and Henry Kucera. 1982. *Frequency analysis of English usage: Lexicon and grammar.* New York: Houghton Mifflin.

Friederici, Angela D. 2010. Brain circuits of syntax. In Derek Bickerton and Eors Szathmary, eds., *Biological foundations and origin of syntax,* 239–252. Cambridge, Mass.: MIT Press.

Frisch, K. von. 1967 Honeybees: Do they use direction and distance information provided by their dances? *Science* 15: 1072–1076.

Futuyma, Douglas. 1983. *Science on trial.* New York: Basic Books.

Ganger, Jennifer, and Michael R. Brent. 2004. Reexamining the vocabulary spurt. *Developmental Psychology* 40(4): 621–632.

Garagnani, Max, Thomas Wennekers, and Friedemann Pulvermüller. 2009. Recruitment and consolidation of cell assemblies for words by way of Hebbian learning and competition in a multi-layer neural network. *Cognitive Computation* 1(2): 160–176.

Gardner, Howard. 1983 Frames of mind: The theory of multiple intelligences. New York: Basic Books.

Gardner, R. A., and B. T. Gardner. 1969. Teaching sign language to a chimpanzee. *Science* 165: 664–672.

Garner, Joseph P., Graham K. Taylor, and Adrian L. R Thomas. 1999. On the origins of birds: the sequence of character acquisition in the evolution of avian flight. *Proceedings of the Royal Society of London. B 22;* 266 (1425): 1259–1266.

Garnham, Alan, Richard V. C. Shillcock, Gordon D. A. Brown, Andrew I. D. Mill, and Ann Cutler. 1981. Slips of the tongue in the London-Lund corpus of spontaneous conversation. *Linguistics* 19: 805–818.

Gathercole, Virginia C. Mueller, and Enlli Mon Thomas. 2009. Bilingual first-language development: Dominant language takeover, threatened minority language take-up. *Bilingualism: Language and Cognition,* 12: 213–237.

Gavrilets, S., and D. Waxman. 2002. Sympatric speciation by sexual conflict. *Proceedings of the National Academy of Science* 99: 10533–10538.

Geist, V. 1987. Commentary on Blumenschine. *Current Anthropology.* 28: 396–397.

Genesee, Fred, and Elena Nicoladis. 2009. Bilingual first language acquisition. In Erika Hoff and Marilyn Shatz, eds. *Blackwell Handbook of Language Development* 324–342. New York: Wiley-Blackwell.

Gentner, Dedre. 1982. Why nouns are learned before verbs: Linguistic relativity versus natural partitioning. Technical Report No. 257. Bolt Beranek and Newman, Inc.

Geschwind, Norman. 1970. The organization of language and the brain. *Science* 27: 940–944.

Gil, D. 1994. The structure of Riau Indonesian. *Nordic Journal of Linguistics* 17: 179–200.

Gilbert, Scott F., and David Epel. 2009. *Ecological developmental biology: Integrating epigenetics, medicine and evolution.* Sunderland, Mass.: Sinauer.

Gilbert, Scott F., and S. Sarkar. 2000. Embracing complexity: Organicism for the twenty-first century. *Developmental Dynamics* 219: 1–9.

Gilligan, Gary. 1987. *A cross-linguistic approach to the pro-drop parameter.* PhD Dissertation, University of Southern California.

Givon, T. 1979. Prolegomena to any sane creology. In Ian Hancock, ed., *Readings in creole studies,* 3–35. Ghent: E Story Scientia.

———. 1987. Universals of discourse structure and second language acquisition. In William Rutherford, ed., *Language universals and second language acquisition,* 109–136. Amsterdam: Benjamins.

———. 1998. Towards a neurology of grammar. *Behavioral and Brain Sciences* 21: 154–155.
Givon, T., and Masayoshi Shibatani, eds. 2009. *Syntactic complexity.* Amsterdam: Benjamins.
Goldfield, B. A., and J. S. Reznick. 1996. Measuring the vocabulary spurt: A reply to Mervis and Bertrand. *Journal of Child Language* 23: 241–246.
Goldin-Meadow, Susan, Martin E. Seligman, and Rochel Gelman. 1976. Language in the two-year-old. *Cognition* 4: 189–202.
Golumbia, David. 2010. Minimalism is functionalism. *Language Sciences* 32: 28–42
Good, Jeff. 2009. A twice-mixed creole? Tracing the history of a prosodic split in the Saramaccan lexicon. *Studies in Language* 33: 459–498.
Goodall, Jane. 1971. *In the shadow of man.* New York: Houghton Mifflin.
Goodman, Corey S., and Bridget Coughlin. 2000. The evolution of evo-devo biology. *Proceedings of the National Academy of Science* 97: 4424–4425.
Goodman, Morris. 1985. Review of Bickerton (1981). *International Journal of American Linguistics* 51: 109–137.
———. 1987. The Portuguese element in the American creoles. In Glenn Gilbert, ed., *Pidgin and creole languages,* 361–405. Honolulu: University of Hawaii Press.
Gould, J. L. 1976. The dance language controversy. *Quarterly Review of Biology* 51: 211–244.
Gould, Stephen J. 1989. *Wonderful life: The Burgess Shale and the nature of history.* New York: Norton.
Griffin, D. 1959. *Listening in the dark.* New Haven, Conn.: Yale University Press.
Griffin, D., and R. Galambos. 1941. The sensory basis of obstacle avoidance by flying bats. *Journal of Experimental Zoology* 86: 481–506.
Griffiths, D., A. Dickinson, and N. Clayton. 1999. Episodic memory: What can animals remember about their past? *Trends in Cognitive Science* 3: 74–80.
Grimshaw, Jane. 1981. Form, function and the language acquisition device. In C. L. Baker and J. J. McCarthy, eds., *The logical problem of language acquisition,* 165–182. Cambridge, Mass.: MIT Press.
Grodzinsky, Yosef, and Angela D. Friederici. 2006. Neuroimaging of syntax and syntactic processing. *Current Opinion in Neurobiology* 16: 240–246.
Gulyas, Balasz. 2010. Functional neuroimaging and the logic of brain operations: Methodologies, caveats and fundamental examples from language research. In Derek Bickerton and Eors Szathmary, eds., *Biological foundations and origin of syntax,* 41–59. Cambridge, Mass.: MIT Press.
Guthrie, L. F. 1984. Contrasts in teachers' language use in a Chinese-English bilingual classroom. In J. Handscombe, F. A. Orem, and B. P. Taylor, eds., *On Tesol '83: The question of control,* 39–42. Washington, D.C.: TESOL.
Hagstrom, Paul. 2001. Course handout, Boston University. http://www.bu.edu/linguistics/UG/course/lx522-f01/handouts/lx522-13-mp.pdf.

Hall, Robert. 1966. *Pidgin and creole languages.* Ithaca, N.Y.: Cornell University Press.
Hamilton, W. D. 1964. The genetical evolution of social behaviour. *Journal of Theoretical Biology* 7: 1–16.
Harnad, Stevan. 2008. Why and how the problem of the evolution of universal grammar (UG) is hard. *Behavioral and Brain Sciences* 3: 524–525.
Harris, Randy Allen. 1993. *The linguistic wars.* Oxford: Oxford University Press.
Harris, Tom. N.d. How spiders work. HowStuffWorks. http://animals.howstuffworks.com/arachnids/spider5.htm.
Hart, Donna, and Robert W. Sussman. 2005. *Man the hunted: Primates, predators and human evolution.* New York: Westview.
Harvey, P. H., and M. D. Pagel. 1991. *The comparative method in evolutionary biology.* Oxford: Oxford University Press.
Haspelmath, Martin. 2008. Parametric versus functional explanations of syntactic universals. In Theresa Biberauer, ed., *The limits of syntactic variation*, 75–107. Amsterdam: Benjamins.
Hauser, Marc D. 1996. *The evolution of communication.* Cambridge, Mass.: MIT Press.
———. 2009. The possibility of impossible cultures. *Nature* 460: 190–196.
Hauser, Marc D., D. Barner, and T. O'Donnell. 2007. Evolutionary linguistics: A new look at an old landscape. *Language Learning and Development* 3: 101–132.
Hauser, Marc D., N. Chomsky, and T. Fitch. 2002. The faculty of language: What is it, who has it, and how did it evolve? *Science* 298: 1569–1579.
Hauser, Marc D., Katherine McAuliffe, and Peter R. Blake. 2009. Evolving the ingredients for reciprocity and spite. *Philosophical Transactions of the Royal Society* B 12, 364: 3255–3266.
Hawkes, K. 1991. Showing off: Tests of an hypothesis about men's foraging goals. *Ethology and Sociobiology* 12: 29–54.
Hayes, Keith J., and Catherine Hayes. 1952. Imitation in a home-raised chimpanzee. *Journal of Comparative and Physiological Psychology* 45: 450–459.
Heath, Scott Christopher, Ruth Schulz, David Ball, and Janet Wiles. 2012. Lingodroids: Learning terms for time. Paper presented at IEEE International Conference on Robotics and Automation, St. Paul, Minn.
Hebb, Donald O. 2002. *The organization of behavior.* New York: Psychology Press.
Hedges, S. B., J. E. Blair, M. L. Venturi, and J. L. Shoe. 2004. A molecular timescale of eukaryote evolution and the rise of complex multicellular life. *BMC Evolutionary Biology* 4: 2.
Heibeck, Tracy H., and Ellen M. Markman. 1987. Word learning in children: An examination of fast mapping. *Child Development* 58: 1021–1034.
Heine, Bernd, and Tania Kuteva. 2007. *The genesis of grammar.* Oxford: Oxford University Press.

---. 2012. Grammaticalization theory as a tool for reconstructing language evolution. In M. Tallerman and K. R. Gibson, eds., *The Oxford handbook of language evolution*, 512–537. Oxford: Oxford University Press.

Heinrich, Bernd. 1991. *Ravens in winter.* London: Vintage.

Henderson, J., T. A. Hurly, M. Bateson, and S. D. Healy. 2006. Timing in free-living rufous hummingbirds *(Selasphorus rufus). Current Biology* 16: 512–515.

Herlein, J. D. 1718. *Beschrijvinge van der volksplantinge Zuriname.* Leeuwarden: Injema.

Herrnstein, R. J. 1985. Riddles of natural categorization. In L. Weiskrantz, ed., *Animal intelligence*, vol. 7, 129–144. Oxford: Clarendon Press.

Hickey, Raymond. 2010. Language change. In Mirjam Fried, Jan-Ola Ostman, and Jef Verschueren, eds., *Variation and change: Pragmatic perspectives*, 171–202. Amsterdam: Benjamins.

Higashi, M., G. Takimoto, and N. Yamamura. 1999. Sympatric speciation by sexual selection. *Nature* 402: 523–526.

Higman, B. W. 1995. *Slave populations of the British Caribbean, 1807–1834.* Kingston, Jamaica: University of the West Indies Press.

Hindman, Hugh D., ed. 2009. *The world of child labor: An historical and regional survey.* Armonk, N.Y.: M. E. Sharpe.

Hobson, J. A. 2009. REM sleep and dreaming: Towards a theory of protoconsciousness. *National Review of Neuroscience* 10: 803–862.

Hockett, Charles F. 1960. The origin of speech. *Scientific American* 203: 89–96.

Hockett, Charles F., and Stuart Altmann. 1968. A note on design features. In Thomas A. Sebeok, ed., *Animal communication: Techniques of study and results of research*, 61–72. Bloomington: Indiana University Press.

Hoekstra, H. E., and J. A. Coyne. 2007. The locus of evolution: Evo devo and the genetics of adaptation. *Evolution* 61: 995–1016.

Holldobler, Bert. 1971. Recruitment behavior in Camponatus socius. Zeitschrift fur Vergleichende Pfysiologie 75: 123–142.

---. 1978. Ethological aspects of chemical communication in ants. *Advances in the Study of Behavior* 8: 75–111.

Holldobler, Bert, and E. O. Wilson. 2009. *The superorganism: The beauty, elegance and strangeness of insect societies.* New York: Norton.

Holm, John. 1986. Substrate diffusion. In P. Muyskem and N. Smith, eds., *Substrate versus universals in creole genesis*, 259–278. Amsterdam: Benjamins.

---. 1988. *Pidgins and Creoles*, vol. 1. Cambridge, U.K.: Cambridge University Press.

Holmberg, Anders, and Urpo Nikanne. 1993. Introduction. In A. Holmberg and U. Nikanne, eds., *Case and other functional categories in Finnish syntax*, 1–22. Berlin: Mouton-deGruyter.

Hopper, Paul J., and Elizabeth Traugott. 2003. *Grammaticalization.* Cambridge, U.K.: Cambridge University Press.

Hornstein, Norbert. 1995. *Logical form: From GB to minimalism*. Oxford: Blackwell.
———. 2001 *Move! A minimalist theory of construal*. Oxford: Blackwell.
———. 2009. *A theory of syntax: Minimal operations and universal grammar*. Cambridge, U.K.: Cambridge University Press.
Huang, C.-T. James. 1984. On the distribution and reference of empty pronouns. *Linguistic Inquiry* 15: 531–574.
Hurford, James. 2003. The neural basis of predicate-argument structure. *Behavioral and Brain Sciences* 26: 261–283.
———. 2007a. *The origin of meaning*. Oxford: Oxford University Press.
———. 2007b. The origin of noun phrases: Reference, truth and communication. *Lingua* 117: 527–542.
Hyams, N. 1986. *Language acquisition and the theory of parameters*. Dordrecht: Reidel.
Isaac, G. 1978. The food-sharing behavior of protohuman hominids. *Scientific American* 238: 90–108.
Jackendoff, Ray. 1977. *X-bar-Syntax: A Study of Phrase Structure*. Linguistic Inquiry Monograph 2. Cambridge, Mass.: MIT Press.
———. 1999. Possible stages in the evolution of the language capacity. *Trends in Cognitive Sciences* 3: 272–279.
———. 2002. *Foundations of language: Brain, meaning, grammar, evolution*. Oxford: Oxford University Press.
James, William. 1890. *The principles of psychology*. London: Macmillan.
Janik, V. M., L. S. Sayigh, and R. S. Wells. 2006. Signature whistle shape conveys identity information to bottlenose dolphins. *Proceedings of the National Academy of Science* 103: 8293–8297.
Jeannerod, M. 1997. *The cognitive neuroscience of action*. Oxford: Blackwell.
Jelinek, Eloise. 1984. Empty categories, case, and configurationality. *Natural Language & Linguistic Theory* 2: 39–76.
Jenkins, Lyle. 2000. *Biolinguistics: Exploring the biology of language*. Cambridge, U.K.: Cambridge University Press.
Johansson, Barbro. 2000. Brain plasticity and stroke rehabilitation: The Willis lecture. *Stroke* 31: 223–230.
Johansson, Sverker. 2005. *Origins of language: Constraints and hypotheses*. London: Allen and Unwin.
Johnson, George. 1995. Chimp talk debate: Is it really language? *New York Times*, June 6.
Johnson, Phillip E. 1997. *Defeating Darwinism by opening minds*. Downers Grove, Ill.: InterVarsity Press.
Jones, D. M. 1956. Review of M. Sandmann. *Classical Review*, new series 6: 184–184.
Jones, Gareth, and Marc W. Holderied. 2007. Bat echolocation calls: Adaptation and convergent evolution. *Proceedings of the Royal Society* B7, 274: 905–912.

Jones, Mari C., and Ishtla Singh. 2005. *Exploring language change*. New York: Routledge.

Joyce, James. 1960. *Ulysses*. London: Bodley Head.

Kaas, Jon H. 1987. The organization of neocortex in mammals: Implications for theories of brain function. *Annual Review of Psychology*, 38: 129–151.

Kandel, Eric R., James H. Schwartz, and Thomas M. Jessell. 2000. *Principles of neural science*, 4th edition. New York: McGraw-Hill.

Kappeler, P. M., and J. B. Silk, eds. 2010. *Mind the gap: Tracing the origin of human universals*. Berlin: Springer Verlag.

Karmiloff-Smith, A. 1992. *Beyond modularity: A developmental perspective on cognitive science*. Cambridge, Mass.: MIT Press.

Kaschube, Matthias, Michael Schnabel, Siegrid Löwel, David M. Coppola, Leonard E. White, and Fred Wolf. 2010. Universality in the evolution of orientation columns in the visual cortex. *Science* 330: 1113–1116.

Katz, Jerrold J. 1964. Mentalism in linguistics. *Language* 40: 124–137.

Kavanau, J. Lee. 2007. Roots of avian evolution: Clues from relict reproductive behaviors. *Scientific Research and Essays* 2: 263–294.

Kawata, M. 2002. Invasion of vacant niches and subsequent sympatric speciation. *Proceedings of the Royal Society, London B* 269: 55–63.

Kayne, Richard S. 1994. The antisymmetry of syntax. *Linguistic Inquiry Monograph 25*. Cambridge, Mass.: MIT Press.

———. 1996. Microparametric syntax: some introductory remarks. In J. Black and V. Motapanyane, eds., *Microparametric syntax and dialect variation*, ix–xvii. Amsterdam: Benjamins.

———. 2005. Some notes on comparative syntax, with special reference to English and French. In Richard S. Kayne and Guglielmo Cinque, eds., *The Oxford Handbook of Comparative Syntax*, 3–69. New York: Oxford University Press.

Kellis, S., K. Miller, K. Thomson, R. Brown, P. House, and B. Greger. 2010. Decoding spoken words using local field potentials recorded from the cortical surface. *Journal of Neural Engineering* 7, no. 5: 056007.

Kenneally, Christine. 2007. *The first word: The search for the origins of language*. New York: Viking.

Kihm, Alain. 2008. The two faces of creole grammar and their implication for the origin of complex language. In R. Eckardt, G. Jager, and T. Veenstra, eds., *Variation, selection, development: Probing the evolutionary model of language change*. 253–306. Berlin: Mouton de Gruyter.

King, Robert D. 1969. *Historical linguistics and generative grammar*. New York: Prentice Hall.

Klima, E. S., and U. Bellugi. 1966. Syntactic regularities in the speech of children. In J. Lyons and R. J. Wales, eds., *Psycholinguistic papers: The proceedings of the 1966 Edinburgh Conference*, 183–219. Edinburgh: Edinburgh University Press.

Klinger, Mark R., Philip C. Burton, and G. Shane Pitts. 2000. Mechanisms of unconscious priming: I. Response competition, not spreading activation. *Journal of Experimental Psychology: Learning, Memory, and Cognition* 26: 441–455.

Kluender, Robert. 2005. Are subject islands subject to a processing account? In Vineeta Chand, Ann Kelleher, Angelo J. Rodríguez and Benjamin Schmeiser, eds., *Proceedings of the 23rd West Coast Conference on Formal Linguistics*, 475–499. Somverville, Mass.: Cascadilla Press

Knowlton, Edgar. 1967. Pidgin English and Portuguese. *Kentucky Foreign Language Quarterly* 7: 212–218.

Kohler, W. 1927. *The mentality of apes.* Trans. E. Winter. New York: Routledge & Kegan Paul.

Koopman, Hilda, and Dominique Sportiche. 1985. The position of subjects. *Lingua* 85: 211–258.

Koopman, Willem F. 1990. *Word order in Old English, with special reference to the verb phrase*, Ph.D.diss., University of Amsterdam.

Koster, Jan. 2007. Structure preservingness, internal merge and the strict locality of triads. In S. Karimi, V. Samiian, and W. K. Wilkins, eds., *Phrasal and clausal architecture: Syntactic derivation and interpretation*, 188–205. Amsterdam: Benjamins.

Kouwenberg, Sylvia. 2009. The demographic context of creolization in early Jamaica 1650–1700. In Rachel Selbach, Hugo C. Cardoso, and Margot van den Berg, eds., *Gradual creolization: Studies celebrating Jacques Arends*, 325–348. Amsterdam: Benjamins.

Kouwenberg, Sylvia, and John Singler. 2008. Introduction. In S. Kouwenberg and J. Singler, eds., *The handbook of pidgin and creole studies*, 1–16. Chichester, U.K.: Wiley-Blackwell.

Kramer, Marvin Gould. 2002. Substrate transfer in Saramaccan Creole. Ph.D. diss., University of California, Berkeley.

Krebs, J. R., and R. Dawkins. 1984. Animal signals: Mind-reading and manipulation, In J.R. Krebs and N.B. Davies, *Behavioural ecology: An evolutionary approach*, 188–205. Oxford: Blackwell.

Kroll, J. F., and J. Curley. 1988. *Practical aspects of memory.* London: Wiley.

Krubitzer, Leah 1995. The organization of neocortex in mammals: Are species differences really so different? *Trends in Neurosciences* 18: 408–417.

Kuhl, P. K. 2010. Brain mechanisms in early language acquisition. *Neuron* 67: 713–727.

Kummer, H. 1971. *Primate societies: Group techniques of ecological adaptation.* Chicago: Aldine.

Labov, William. 1972. *Sociolinguistic patterns.* Philadelphia: University of Pennsylvania Press.

Laland, K. N., F. J. Odling-Smee, and M. W. Feldman. 1999. Evolutionary consequences of niche construction and their implications for ecology. *Proceedings of the National Academy of Sciences* 96: 10242–10247.

Laland, K. N., J. Odling-Smee, and S. F. Gilbert. 2008. Evo-devo and niche construction: Building bridges. *Journal of Experimental Zoology* 310B: 549–566.

Laland, K. N., and Kim Sterelny. 2006. Seven reasons (not) to neglect niche construction. *Evolution* 60: 1751–1762.

Lang, George. 2008. *Making Wawa: The genesis of Chinook jargon.* Vancouver: UBC Press,

Larick, R., and R. Ciochan. 1996. The African emergence and early Asian dispersals of the genus Homo. *American Scientist* 84: 538–551.

Larson, Richard K. 1988. On the double-object construction. *Linguistic Inquiry* 19: 335–391.

Larson, Richard K., V. Deprez, and H. Yamakido, eds. 2010. *The evolution of human language: Biolinguistic perspectives.* Cambridge, U.K.: Cambridge University Press.

Lasnik, Howard, and Juan Uriagereka. 1988. *A course in GB syntax: Lectures on binding and empty categories.* Cambridge, Mass.: MIT Press.

Lau, E. F., and F. Ferreira. 2005. Lingering effects of disfluent material on comprehension of garden path sentences. *Language and Cognitive Processes* 20: 633–666.

Lefebvre, Claire. 1986. Relexification in creole genesis revisited: The case of Haitian Creole. In P. Muysken and N. Smith, eds., *Substrata versus universals in creole language genesis,* 279–301. Amsterdam: Benjamins.

———. 1995. The role of relexification and syntactic reanalysis in Haitian Creole: Methodological aspects of a research program. In S. Mufwene, ed., *Africanisms in Afro-American language varieties,* 254–279. Athens: University of Georgia Press.

———. 1999. *Creole genesis and the acquisition of grammar.* Cambridge, U.K.: Cambridge University Press.

Lehmann, J., A. H. Korstjens, and R. I. M. Dunbar. 2007. Fission-fusion social systems as a strategy for coping with ecological constraints: A primate case. *Evolutionary Ecology,* 21: 613–634.

Li, Gang, Jinhong Wang, Stephen J. Rossiter, Gareth Jones, and Shuyi Zhang. 2007. Accelerated FoxP2 evolution in echolocating bats. *PloS ONE* 2(9): e900.

Lieberman, Philip. 1984. *The biology and evolution of language.* Cambridge, Mass.: Harvard University Press.

———. 2010. The creative capacity of language, in what manner is it unique, and who had it? In R. K. Larson, V. Deprez, and H. Yamakido, eds., *The evolution of human language: Biolinguistic perspectives,* 163–175. Cambridge, U.K.: Cambridge University Press.

Lightfoot, David. 1991a. *How to set parameters: Arguments from language change.* Cambridge, Mass.: MIT Press.

———. 1991b. Subjacency and sex. *Language and Communication* 11: 67–69.

Liu, Y, J. A. Cotton, B. Shen, X. Han, S. J. Rossiter, and S. Zhang. 2010. Convergent sequence evolution between echolocating bats and dolphins. *Current Biology* 20: R53–54.

Long, M. 1996. The role of the linguistic environment in second language acquisition. In W. C. Ritchie and T. K. Bhatia, eds., *Handbook of second language acquisition*, 413–468. New York: Academic Press.

Longa, Victor M., and Guilermo Lorenzo. 2008. What about a (really) minimalist theory of language acquisition? *Linguistics* 46: 541–570,

Lord, Carol. 1973. Serial verbs in transition. *Studies in African Linguistics* 4: 269–296.

Lorenzo, Guillermo. 2012. The evolution of the faculty of language. In Cedric Boeckx, Maria del Carmen Horno-Cheliz, and Jose-Luis Mendivil-Giro, eds., *Language, from a Biological Point of View: Current Issues in Biolinguistics*, 263–289. Newcastle upon Tyne, U.K.: Cambridge Scholars Publishing.

Lust, Barbara. 1986. Introduction. In B. Lust, ed., *Studies in the acquisition of anaphora:* vol. 1, *Defining the constraints*, 3–103. Dordrecht: Reidel.

———. 1999. Universal grammar: The strong continuity hypothesis in first language acquisition. In William C. Ritchie and Tej K. Bhatia, eds., *Handbook of language acquisition*, 111–155. San Diego, Calif.: Academic Press.

MacWhinney, Brian. 1976. Hungarian research on the acquisition of morphology and syntax. *Journal of Child Language* 3: 397–410.

Manley, Geoffrey, and Jennifer Clack. 2004. An outline of the evolution of vertebrate hearing organs. In Geoffrey A. Manley, A. N. Popper, and R. R. Faye, eds., *Evolution of the vertebrate auditory system*, 1–26. New York: Springer.

Marcus, Gary. 2004. *The birth of the mind*. New York: Basic Books.

Marler, Peter. 1991. The instinct to learn. In S. Carey and E. Gelman, eds., *The epigenesist of mind: Essays in biology and cognition*, 37–66. Hillsdale, N.J.: Lawrence Erlbaum.

Marler, Peter, and Douglas A. Nelson. 1992. Action-based learning: A new form of developmental plasticity in bird song. *Netherlands Journal of Zoology* 43: 91–103.

Marler, Peter, and S. Peters. 1982. Subsong and plastic song: Their role in the vocal learning process. In D. E. Kroodsma and E. H. Miller, eds., *Acoustic communication in birds*, vol. 2, 25–50. New York: Academic Press.

Martin, Alex. 2007. The representation of object concepts in the brain. *Annual Review of Psychology* 58: 25–45.

Mather, Patrick-Andre. 2006. Second language acquisition and creolization: Same (i-) processes, different (e-) results. *Journal of Pidgin and Creole Languages* 21: 231–274.

May, Robert. 1977. The grammar of quantification. Ph.D. diss., Massachusetts Institute of Technology.

Maynard Smith, John. 1966. Sympatric speciation. *American Naturalist* 100: 637–650.

———. 1982. *Evolution and the theory of games.* Cambridge, U.K.: Cambridge University Press.

Maynard Smith, John, and G. R. Price. 1973. The logic of animal conflict. *Nature* 246: 15–18.

Maynard Smith, John, and Eors Szathmary. 1995. *The major transitions in evolution.* Oxford: Freeman.

Mayr, Ernst. 1963. *Animal Species and Evolution.* Cambridge, Mass.: Belknap Press of Harvard University Press

———. 1995. Can SETI succeed? Not likely. *Bioastronomy News* 7: 3.

McDaniel, Dana. 2005. The potential role of production in the evolution of syntax. In Maggie Tallerman, ed., *Language origins: Perspectives on evolution,* 153–165. Oxford: Oxford University Press.

McGhee, George. 2011. *Convergent evolution: Limited forms most beautiful.* Cambridge, Mass.: MIT Press.

McWhorter, John. 1992. Substratal influence in Saramaccan serial verb constructions. *Journal of Pidgin and Creole Languages* 7: 1–53.

———. 1994. Rejoinder to Derek Bickerton. *Journal of Pidgin and Creole Languages* 9: 79–93.

———. 1997. *Towards a new model of creole genesis.* New York: Peter Lang.

———. 2008. Inflectional morphology and universal grammar: Post hoc versus propter hoc. In R. Eckardt, G. Jager, and T. Veenstra, eds., *Variation, selection, development: Probing the evolutionary model of language change,* 337–373. Berlin: Mouton de Gruyter.

———. 2013. Language Turned Off?: The legacy of the bioprogram hypothesis. *Journal of Pidgin and Creole Languages.* 28: 131–136.

Megyesi, Beata. N.d. The Hungarian language: A short descriptive grammar. http://stp.ling.uu.se/~bea/publ/megyesi-hungarian.pdf.

Meisel, Jurgen M. 1989. Early differentiation of language in bilingual children. In Kenneth Hyltenstam and Loraine K. Obler, eds., *Bilingualism across the lifespan: Aspects of acquisition, maturity and loss,* 13–40. Cambridge, U.K.: Cambridge University Press.

———. 2004. The bilingual child. In Tej K. Bhaktia and Willam C. Ritchie, eds., *The handbook of bilingualism,* 94–113. Malden, Mass.: Blackwell.

Mesthrie, Rjend. 1989. The origins of Fanagalo. *Journal of Pidgin and Creole Languages* 4: 211–240.

Migge, Bettina. 1998. Substrate influence in creole formation: The origin of *give*-type serial verb constructions in the Surinamese plantation creole. *Journal of Pidgin and Creole Languages* 13: 215–265.

Miles, H. L. 1990. The cognitive foundations for reference in a signing orangutan. In S. T. Parker and K. R. Gibson, eds., *"Language" and intelligence in monkeys and apes: Comparative developmental perspectives*, 511–539. Cambridge, U.K.: Cambridge University Press.

———. 1994. Me Chantek: The development of self-awareness in a signing orangutan. In S.T. Parker, R.W. Mitchell, and M.L. Boccia, eds., *Self-awareness in animals and humans: Developmental perspectives*, 254–272. New York: Cambridge University Press.

Miller, George A. 1962. Some psychological studies of grammar. *American Psychologist* 17: 748–762.

———. 1975. Some comments on competence and performance. *Annals of the New York Academy of Sciences* 263: 201–204.

Mills, Annie E. 1985. The acquisition of German. In Dan Slobin, ed., *The crosslinguistic study of acquisition data:* vol. 1, *The data*, 141–254. Hillsdale, N.J.: Lawrence Erlbaum.

Mitani, John C. 1996. Comparative studies of African ape vocal behavior. In William C. McGrew, Linda F. Marchant, and Toshisada Nishida, eds., *Great ape societies*, 241–254. Cambridge, U.K.: Cambridge University Press.

Mitani, John, and D. Watts. 1999. Demographic influences on the hunting behavior of chimpanzees. *American Journal of Physical Anthropology* 109: 439–454.

Mithen, Steven J. 1996. *The prehistory of the mind: A search for the origins of art, religion, and science*, London: Thames and Hudson

———. 2005. *The singing Neanderthals: The origin of language, music, mind and body*. London: Weidenfeld and Nicholson.

Moglich, M., and B. Holldobler. 1975. Communication and orientation during foraging and emigration in the ant *Formica fusca*. *Journal of Comparative Physiology* 101: 275–288.

Monahan, C. M. 1996. New zooarchaeological data from Bed II, Olduvai Gorge, Tanzania: Implications for hominid behavior in the Early Pleistocene. *Journal of Human Evolution* 31: 93–128.

Mtui, Estomih, and Gregory Gruener. 2006. *Clinical neuroanatomy and neuroscience*. Philadelphia: Saunders.

Mufwene, Salikoko. 1996a. The development of American Englishes: Some questions from a creole genesis perspective. In Edgar W. Schneider, ed., *Focus on the USA: Varieties of English around the world*, 231–264. Amsterdam: Benjamins.

———. 1996b. The founder principle in creole genesis. *Diachronica* 13: 83–134.

———. 2001. *The ecology of language evolution*. Cambridge, U.K.: Cambridge University Press.

———. 2003. Genetic linguistics and genetic creolistics: A response to Sarah G. Thomason. *Journal of Pidgin and Creole Languages* 18: 273–288.

———. 2008. What do creoles and pidgins tell us about the evolution of language? In Bernard Laks, ed., *Origin and evolution of languages*, 272–297. Sheffield, U.K.: Equinox.

———. 2009. The evolution of language: Hints from pidgins and creoles. In James Minett and William Wang, eds., *Language, evolution, and the brain,* 1–33. Hong Kong: City University of Hong Kong Press.

Muysken, Pieter. 1988. Are Creoles a special type of language? In Frederick Newmeyer, ed., *Linguistics: The Cambridge survey:* vol. 2, *Linguistic theory: Extensions and implications,* 285–301. Cambridge, U.K.: Cambridge University Press.

Nazzi, Nadarajah B., J. Brunstrom, J. Grutzendler, R. Wong, and A. Pearlman. 2001. Two modes of radial migration in early development of the cerebral cortex. *Nature Neuroscience* 4: 143–150.

Nazzi, Thierry, and Josiane Bertoncini. 2003. Before and after the vocabulary spurt: Two modes of word acquisition? *Developmental Science* 6: 136–142.

Nelson, Diane Carlita. 1998. *Grammatical case assignment in Finnish.* New York: Garland.

Nelson, Katherine. 1981. General discussion. In Harold Winitz, ed., *On native language and foreign language acquisition. Annals of the New York Academy of Science* 379: 215.

Neuweiler, Gerhard. 2003. Evolutionary aspects of bat echolocation. *Journal of Comparative Physiology A* 189: 245–256.

Newell, Alan. 1992. SOAR as a unified theory of cognition: Issues and explanations. *Behavioral and Brain Sciences* 15: 462–495.

Newen, Albert, and Andreas Bartels. 2007. Animal minds and the possession of concepts. *Philosophical Psychology* 20: 283–308.

Newmeyer, Frederick J. 1993. Chomsky's philosophy of language. In R. E. Asher and J. M. Y. Simpson, eds. *The encyclopedia of language and linguistics,* 536–541. Oxford: Pergamon Press.

———. 1996. *Generative linguistics: A historical perspective.* New York: Routledge.

———. 2004. Against a parameter-setting approach to language variation. In Pierre Pica, Johan Rooryck, and J. van Craenenbroek, eds., *Language Variation Yearbook, Vol. 4,* 181–234. Amsterdam: Benjamins.

———. 2005. *Possible and probable languages: A generative perspective on linguistic typology.* Oxford: Oxford University Press.

Newport, E. L., H. Gleitman, and L. R. Gleitman. 1977. Mother, I'd rather do it myself: Some effects and non-effects of maternal speech style. In C. E. Snow and C. A. Ferguson, eds., *Talking to children: Language input and acquisition,* 109–149. Cambridge, U.K.: Cambridge University Press.

Newton, Michael. 2002. *Savage girls and wild boys: A history of feral children.* London: Faber and Faber.

Nishihara, Hidenori, Yoko Satta, Masato Nikaido, J. G. M. Thewissen, Michael J. Stanhope, and Norihiro Okada. 2005. A retroposon analysis of Afrotherian phylogeny. *Molecular Biology and Evolution* 22: 1823–1833.

Nottebohm, F. 2005. The neural basis of birdsong. *PLOS Biology* 3, no. 5: e164.
Novakowski, N. S. 1967. The winter bioenergetics of a beaver population in northern latitudes. *Canadian Journal of Zoology* 45: 1107–1118.
Nowak, Martin A., Corina E. Tarnita, and Edward O. Wilson. 2010. The evolution of eusociality. *Nature* 466: 1057–1062.
O'Brien, E. 1981. The projectile capabilities of an Acheulian handaxe from Olorgesailie. *Current Anthropology* 22: 76–79.
Odling-Smee, F. J., K. N. Laland, and N. W. Feldman. 2003. *Niche construction: The neglected process in evolution.* Princeton, N.J.: Princeton University Press.
O'Grady, William. 1987. *Principles of grammar and learning.* Chicago: University of Chicago Press.
Ojemann, G. A. 1991. Cortical organization of language. *Journal of Neuroscience* 11: 2281–2287.
Owren, Michael J., Drew Rendall, and Michael J. Ryan. 2010. Redefining animal symbolism: Influence versus information in communication. *Biology and Philosophy* 25: 755–780.
Papen, Robert. 1978. The French-based creoles of the Indian Ocean: An analysis and comparison. Ph.D. diss., University of California, San Diego.
Park, M. S., A. D. Nguyen, H. E. Aryan, U. Hoi Sang, M. L. Levy, and K. Semendeferi. 2007. Evolution of the human brain: Changing brain size and the fossil record. *Neurosurgery* 60: 555–562.
Passingham, R. E. 1975. Changes in the size and organisation of the brain in man and his ancestors. *Brain, Behavior and Evolution* 11: 73–90.
Patterson, Francine. 1978. Conversations with a gorilla. *National Geographic* 154: 438–465.
Patterson, Karalyn, Peter J. Nestor, and Timothy T. Rogers. 2007. Where do you know what you know? The representation of semantic knowledge in the human brain. *Nature Reviews/Neuroscience* 8: 976–987.
Paus, T., L. Petrides, A. C. Evans, and E. Meyer. 1993. Role of the human anterior cingulate cortex in the control of oculomotor, manual, and speech responses: A positron emission tomography study. *Journal of Neurophysiology* 70: 453–469.
Pearce, J. M. 1989. The acquisition of an artificial category by pigeons. *Quarterly Journal of Experimental Psychology* 41B: 381–406.
Penn, Derek C., Keith Holyoak, and Daniel Povinelli. 2008. Darwin's mistake: Explaining the discontinuity between human and nonhuman minds. *Behavioral and Brain Sciences* 31: 109–178.
Pepperberg, Irene M. 1991. *The Alex studies: Cognitive and communicative abilities of grey parrots.* Cambridge, Mass.: MIT Press.
———. 2005. Intelligence and rationality in parrots. In S. Hurley and M. Nudds, *Rational animals?* 469–488. Oxford, U.K.: Oxford University Press.

———. 2007. Emergence of linguistic communication: Studies on grey parrots. In Caroline Lyon, Chrystopher L. Nehaniv, and Angelo Cangelosi, eds., *Emergence of communication and language*, 355–386. London: Springer-Verlag.

Pesetsky, David, and Esther Torrego. 2006. Probes, goals, and syntactic categories. In Y. Otsu, ed., *Proceedings of the 7th Annual Conference on Psycholinguistics*, 1–37. Keio, Japan: Keio University.

Peters, A. M. 1983. *Units of language acquisition*. Cambridge, U.K.: Cambridge University Press.

Petersen, J. 1988 Word-internal code-switching constraints in a bilingual child's grammar. *Linguistics*, 26: 479–493

Petersen, S. E., P. T. Fox, A. Z. Snyder, and M. E. Raichle. 1990. Activation of extrastriate and frontal cortical areas by visual words and word-like stimuli. *Science* 249: 1041–1044.

Piattelli-Palmarini, Massimo. 1989. Evolution, selection, and cognition: From "learning" to parameter setting in biology and the study of language. *Cognition* 31: 1–44.

———. 2010. What is language, that it may have evolved, and what is evolution, that it may apply to language? In R.K. Larson, V. Deprez, and H. Yamakido, eds., *The evolution of human language: Biolinguistic perspectives*, 148–162. Cambridge, U.K.: Cambridge University Press.

Piattelli-Palmarini, Massimo, Roeland Hancock, and Thomas Bever. 2008. Language as ergonomic perfection. *Behavioral and Brain Sciences* 31: 530–531.

Pickering, Martin, and Guy Barry. 1991. Sentence processing without empty categories. *Cognitive Processes* 6: 229–259.

Pickering, Travis Rayne, Charles P. Egeland, Manuel Domínguez-Rodrigo, C. K. Brain, and Amy G. Schnell. 2007. Testing the "shift in the balance of power" hypothesis at Swartkrans, South Africa: Hominid cave use and subsistence behavior in the Early Pleistocene. *Journal of Anthropological Archaeology* 27: 30–45.

Pierce, Amy. 1989. On the emergence of syntax: A cross-linguistic study. Ph.D. diss., Massachusetts Institute of Technology.

Piersma, Thomas, and Jan van Gils. 2010. *The flexible phenotype: A body-centred integration of ecology, physiology and behavior*. Oxford: Oxford University Press.

Pinker, Steven. 1984. *Language learnability and language development*. Cambridge, Mass.: Harvard University Press.

———. 1989. *Learnability and cognition: The acquisition of verb-argument structure*. Cambridge, Mass.: Harvard University Press.

———. 1994. *The language instinct*. New York: Harper Collins.

———. 1995. Language acquisition. In L. R. Gleitman, M. Liberman, and D. N. Osherson, eds., *An invitation to cognitive science:* vol. 1: *Language*, 2nd edition, 135–182. Cambridge, Mass.: MIT Press.

Pinker, Steven, and Paul Bloom. 1990. Natural language and natural selection. *Behavioral and Brain Sciences* 13: 707–784.

Pintzuk, Susan. 1996. Old English verb-complement word order and the change from OV to VO. *York Papers in Linguistics,* No. 17. York, U.K.: University of York.

———. 1999. *Phrase structures in competition: Variation and change in Old English word order.* New York: Garland.

Plag, Ingo. 1995. The emergence of *taki* as a complementizer in Sranan: On substrate influence, universals and gradual creolization. In Jacques Arends, ed., *The early stages of creolization,* 113–148. Amsterdam: Benjamins.

———. 2008a. Creoles as interlanguages: Inflectional morphology. *Journal of Pidgin and Creole Languages* 23: 109–130.

——— 2008b. Creoles as interlanguages: Syntactic structures. *Journal of Pidgin and Creole Languages* 23: 307–328.

Poldrack, R. A., F. W. Sabb, K. Foerde, S. M. Tom, R. E., Asarnow, S. Y. Bookheimer, and B. J. Knowlton. 2005. The neural correlates of motor skill automaticity. *Journal of Neuroscience* 25: 5356–5364.

Pollick, A. S., and F. B. M. de Waal. 2007. Ape gestures and language evolution. *Proceedings of the National Academy of Science* 104: 8184–8189.

Pollock, Jean-Yves. 1989. Verb movement, universal grammar, and the structure of IP. *Linguistic Inquiry* 20: 365–424.

Postma, Johannes. 1990. *The Dutch in the Atlantic slave trade, 1600–1815.* Cambridge, U.K.: Cambridge University Press.

Premack, David. 1986, *Gavagai!: or the Future History of the Animal Language Controversy.* Cambridge, Mass.: MIT Press.

Price, C. J., R. J. S. Wise, and R. S. J. Frackowiak. 1996. Demonstrating the implicit processing of visually presented words and pseudowords. *Cerebral Cortex* 6: 62–70.

Price, Richard. 1976. *The Guiana Maroons.* Baltimore, Md.: Johns Hopkins University Press.

Progovac, Ljiljana. 2009. Sex and syntax: Subjacency revisited. *Biolinguistics* 3: 305–336.

Pulvermuller, Friedemann. 1999. Words in the brain's language. *Behavioral and Brain Sciences* 22: 253–336.

———. 2001. Brain reflections of words and their meaning. *Trends in Cognitive Science* 5: 517–524.

———. 2002. *The neuroscience of language: On brain circuits of word and serial order.* Cambridge, U.K.: Cambridge University Press.

———. 2008. Brain-embodiment of category-specific semantic memory circuits. In Gun T. Semin and Eliot R. Smith, eds., *Embodied grounding: Social, cognitive, affective and neuroscientific approaches,* 71–97. Cambridge, U.K.: Cambridge University Press.

Pylyshyn, Zenon W. 1973. The role of competence theories in cognitive psychology. *Journal of Psycholinguistic Research* 2: 21–50.

Raby, C. R., D. M. Alexis, A. Dickinson, and N. S. Clayton. 2007. Planning for the future by western scrub-jays. *Nature* 445: 919–921.
Radford, Andrew. 1986. Small children's small clauses. *Research Papers in Linguistics* 1: 1–44. (University College of North Wales, Bangor.)
———. 1990. *Syntactic theory and the acquisition of English syntax: The nature of early child grammars of English*. Oxford: Blackwell.
Ralls, Katherine, Patricia Fiorelli, and Sheri Gish. 1985. Vocalizations and vocal mimicry in captive harbor seals, *Phoca vitulina*. *Canadian Journal of Zoology* 63: 1050–1056.
Rat Genome Sequencing Project Consortium. 2004. Genome sequence of the Brown Norway Rat yields insights into mammalian evolution. *Nature* 428: 493–521.
Reed, Kaye E. 1997. Early hominid evolution and ecological change through the African Plio-Pleistocene. *Journal of Human Evolution* 32(2–3): 289–322
Reinecke, John E. 1969. *Language and dialect in Hawaii: A sociolinguistic history to 1935*. Honolulu: University of Hawaii Press.
———. 1997. *The Filipino piecemeal sugar strike of 1924–1925*. Honolulu: University of Hawaii Press.
Reinecke, J. E., and A. Tokimasa. 1934. The English dialect of Hawaii. *American Speech* 9: 48–58.
Rendell, Drew, Michael J. Owren, and Michael J. Ryan. 2009. What do animal signals mean? *Animal Behaviour* 78: 233–240.
Richard, P. B. 1983. Mechanisms and adaptation in the constructive behaviour of the beaver *(C. fiber L.)*. *Acta Zoologica Fennica* 174: 105–108.
Richards, B. J. 1990. *Language development and individual differences: A study of auxiliary verb learning*. Cambridge, U.K.: Cambridge University Press.
Richards, D. G., J. P. Wolz, and L. M. Herman. 1984. Vocal mimicry of computer-generated sounds and vocal labeling of objects by a bottlenosed dolphin, *Tursiops truncates*. *Journal of Comparative Psychology* 98: 10–28.
Richardson, Bonham C., ed. 1992. *The Caribbean in the wider world*. Cambridge, U.K.: Cambridge University Press.
Riddoch, M. J., G. W. Humphreys, M. Coltheart, and E. Funnel. 1988. Semantic systems or system? Neuropsychological evidence re-examined. *Cognitive Neuropsychology* 5: 3–25.
Rilling, James K., Matthew F. Glasser, Todd M. Preuss, Xiangyang Ma, Tiejun Zhao, Xiaoping Hu, and Timothy E. J. Behrens. 2008. The evolution of the arcuate fasciculus revealed with comparative DTI. *Nature Neuroscience* 11: 426–428.
Ristau, Carolyn. 1999. Cognitive ethology. In R. K. Wilson and F. C. Keil, eds., *MIT encyclopedia of the cognitive sciences*, 132–134. Cambridge, Mass.: MIT Press.

Roberts, Ian. 1999. Verb movement and markedness. In M. DeGraff, ed., *Language creation and language change: Creolization, diachrony and development*, 287–328. Cambridge, Mass.: MIT Press.

Roberts, Ian, and Anders Holmberg. 2005. On the role of parameters in universal grammar: A reply to Newmeyer. In H. Brockhuis, ed., *Organizing grammar*, 538–553. Berlin: de Gruyter.

Roberts, S. J. 1995. Pidgin Hawaiian: A sociohistorical study. *Journal of Pidgin and Creole Languages* 10: 1–56.

———. 1998. The role of diffusion in the genesis of Hawaiian Creole. *Language* 74: 1–39.

———. 2000. Nativization and the genesis of Hawaiian Creole. In J. McWhorter, ed., *Language change and language contact in pidgins and creoles*, 257–300. Amsterdam: Benjamins.

———. 2004. The emergence of Hawaii Creole English in the early twentieth century: The sociohistorical context of creole genesis. Ph.D. diss., Stanford University.

———. N.d. Language in Hawaii from 1789 to 1960: The documentary sources. Unpublished manuscript, University of Hawaii.

Robertson, I. H., and J. M. Murre. 1999. Rehabilitation of brain damage: Brain plasticity and principles of guided recovery. *Psychological Bulletin* 125: 544–575.

Ross, John Robert. 1967. Constraints on variables in syntax. Ph.D. diss., Massachusetts Institute of Technology.

Rounds, Carol. 2009. *Hungarian, an essential grammar.* New York: Routledge.

Roverud, R. C. 1993. Neural computations for sound pattern recognition: Evidence for summation of an array of frequency filters in an echolocating bat. *Journal of Neuroscience* 13: 2306–2312.

Rumbaugh, Duane M., and Timothy V. Gill. 1976. The mastery of language-type skills by the chimpanzee *(Pan)*. *Annals of the New York Academy of Sciences* 280: 562–578.

Saffran, E. M., and M. F. Schwartz. 1994. Of cabbages and things: Semantic memory from a neuropsychological perspective. In C. Umilta and M. Moskovitch, eds., *Attention and performance 15*, 507–534. Hillsdale, N.J.: Lawrence Erlbaum.

Saffran, J. R., R. N. Aslin, and E. L. Newport. 1996. Statistical learning by 8-month old infants. *Science* 274: 1926–1928.

Sapir, Edward. 1921. *Language: An introduction to the study of speech.* New York: Harcourt Brace.

Sasaki, Yuki, Wim Vanduffel, Tamara Knutsen, Christopher Tyler, and Roger Tootell. 2005. Symmetry activates extrastriate visual cortex in human and nonhuman primates. *Proceedings of the National Academy of Science* 102: 3159–3163.

Saussure, Ferdinand de. 1959. *Course in general linguistics.* Trans. Wade Baskin. New York: Philosophical Library.

Savage-Rumbaugh, E. S., and William M. Fields. 2000. Linguistic, cultural and cognitive capacities of bonobos *(Pan Paniscus)*. *Culture and Psychology* 6: 131–153.

Savage-Rumbaugh, E. S., J. Murphy, R. A. Sevcik, K. E. Brakke, S. L. Williams, L. Williams, D. M. Rumbaugh, and E. Bates. 1993. Language comprehension in ape and child. *Monographs of the Society for Research in Child Development* 88, no. 3/4.

Savage-Rumbaugh, E. S., and D. M. Rumbaugh. 1993. The emergence of language. In K. R. Gibson and T. Ingold, eds., *Tools, language and cognition in human evolution*, 86–108. Cambridge, U.K.: Cambridge University Press.

Schick, K.D., and N. Toth. 1993. *Making silent stones speak: Human evolution and the dawn of technology*. New York: Simon and Schuster.

Schiefflin, Bambi. 1985. The acquisition of Kaluli. In Dan Slobin, ed., *The crosslinguistic study of acquisition data:* vol. 1, *The data,* 525–593. Hillsdale, N.J.: Lawrence Erlbaum.

Schino, Gabriele, and Filippo Aureli. 2008. Grooming reciprocation among female primates: A meta-analysis. *Biology Letters* 4: 9–11.

Schoenemann, P. Thomas. 2009. Evolution of brain and language. *Language Learning* 59 (Suppl. 1): 162–196

Schuchardt, Hugo. 1980. Die sprache der Saramakkaneger in Surinam. In Glenn Gilbert, ed., *Pidgin and creole languages: Selected essays by Hugo Schuchardt*, 89–126. Cambridge, U.K.: Cambridge University Press.

Schulz, Ruth, Arren Glover, Michael J. Milford, Gordon Wyeth, and Janet Wiles. 2011. Lingodroids: Studies in spatial cognition and language. Paper presented at IEEE International Conference on Robotics and Automation, Shanghai, China, May 9–11.

Scollon, Ron. 1976. *Conversations with a one year old: A case study of the developmental foundation of syntax.* Honolulu: University of Hawaii Press,

Scotland, Robert W. 2010. Deep homology: A view from systematics. *BioEssays* 32: 438–449.

Sebba, Mark. 1987. *The syntax of serial verbs.* Amsterdam: Benjamins.

Seely, T. Daniel 2006, Merge, derivational c-command, and subcategorization in a label-free syntax. In Cedric Boeckx, ed., *Minimalist essays,* 182–218. Amsterdam: Benjamins.

Seyfarth, Robert M., and Dorothy L. Cheney. 1984. Grooming, alliances and reciprocal altruism in vervet monkeys. *Nature* 308: 541–543.

———. 1990. *How monkeys see the world.* Chicago: University of Chicago Press.

Shettleworth, Sara J. 2010. *Cognition, evolution and behavior.* 2nd edition. Oxford: Oxford University Press.

Shubin, Neil, Cliff Tabin, and Sean Carroll. 2009. Deep homology and the origins of evolutionary novelty. *Nature* 457: 818–823.

Siegel, Jeff. 2000. Substrate influence in Hawai'i Creole English. *Language in Society* 29: 197–236.

References

———. 2007. Recent evidence against the language bioprogram hypothesis: The pivotal case of Hawai'i Creole. *Studies in Language* 31: 51–88.

———. 2009. Language contact and second language acquisition. In William C. Ritchie and Tej K Bhatia, eds., *The new handbook of second language acquisition*, 569–589. Bingley, U.K.: Emerald.

Simons, D. J., and M. S. Ambinder. 2005. Change blindness: Theory and consequences. *Current Directions in Psychological Science* 14: 44–48.

Simons, D. J., and R. Rensink. 2005. Change blindness: Past, present and future. *Trends in Cognitive Sciences* 9: 16–20.

Simpson, George G. 1964. The nonprevalence of humanoids. *Science* 143: 769

Singler, John V. 1986. Short note. *Journal of Pidgin and Creole Languages* 1: 141–145.

———. 2006. Children and creole genesis. *Journal of Pidgin and Creole Languages* 21: 157–173.

Slobin, Dan L. 1973. Cognitive prerequisites for the development of grammar. In C. A. Ferguson and D. I. Slobin, eds., *Studies of child language development*, 407–431. New York: Holt, Rinehart and Winston.

———, ed. 1985. *The crosslinguistic study of acquisition data:* vol. 1, *The data*. Hillsdale, N.J.: Lawrence Erlbaum.

———. 2004. From ontogenesis to phylogenesis: What can child language tell us about language evolution? In J. Langer, S. T. Parker, and C. Milbrath, eds., *Biology and knowledge revisited: From neurogenesis to psychogenesis*, 255–285. Mahwah, N.J.: Lawrence Erlbaum.

Smith, Neil. 2000. Foreword. In N. Smith, ed., *Chomsky: New horizons in the study of language and mind*, vi–xvi. Cambridge, U.K.: Cambridge University Press.

Smith, Norval. 1987. The genesis of the creole languages of Surinam. PhD diss., University of Amsterdam.

———. 2006. Very rapid creolization in the framework of the restricted motivation hypothesis. In C. Lefebvre, L. White, and C. Jourdan, eds., *L2 acquisition and creole genesis*, 49–65. Amsterdam: Benjamins.

———. 2008. The origin of the Portuguese words in Saramaccan. In Susanne Michaelis, ed., *Roots of creole structures: Weighing the contribution of substrates and superstrates*, 153–168. Amsterdam: Benjamins.

Son, Minjeong, and Peter Svenonius. 2008. Microparameters of cross-linguistic variation: Directed motion and resultatives. In Natasha Abner and Jason Bishop, eds., *Proccedings of the 27th West Coast Conference on Formal Linguistics*, 388–396. Somerville, Mass.: Cascadilla Press.

Stanhope, M. J., V. G. Waddell, O. Madsen, W. W. de Jong, S. B. Hedges, G. C. Cleven, D. Kao, and M. S. Springer. 1998. Molecular evidence for multiple origins of Insectivora and for a new order of endemic African insectivore mammals. *Proceedings of the National Academy of Sciences* 95: 9967–9972.

Spencer, Andrew. 2008. Does Hungarian have a case system? In Greville C. Corbett and Michael Noonan, eds., *Case and grammatical relations*, 35–56. Amsterdam: Benjamins.

Spoor, F., M. G. Leakey, P. N. Gathogo, F. H. Brown, S. C. Antón, I. McDougall, C. Kiarie, F. K. Manthi, and L. N. Leakey. 2007. Implications of new early Homo fossils from Ileret, east of Lake Turkana, Kenya. *Nature* 448: 688–691.

Stephens, D. W., and J. R. Krebs. 1986. *Foraging theory*. Princeton, N.J.: Princeton University Press.

Stockwell, Robert P., and Donka Minkova. 1991. Subordination and word order change in the history of English. In Dieter Kastovsky, ed., *Historical English syntax*, 367–408. New York: Mouton de Gruyter.

Stringer, C., and C. Gamble. 1993. *In search of the Neanderthals*. London: Thames and Hudson.

Summers, Robert W., and Melvin K. Neville. 1978. On the sympatry of early hominids. *American Anthropologist* 80: 657–660.

Syea, Anand. 2013. Serial verb constructions in Indian Ocean French Creoles (IOCs): Substrate, universal, or an independent diachronic development? *Journal of Pidgin and Creole Languages* 28: 13–64.

Szamado, S., and E. Szathmary. 2006. Selective scenarios for the emergence of natural language. *Trends in Ecology and Evolution* 21: 555–561.

Szathmary, Eors. 2001. Origin of the human language faculty: The language amoeba hypothesis. In J. Trabant and D. S. Ward, eds., *New essays on the origin of language*, 41–51. New York: Mouton de Gruyter.

Tallerman, Maggie. 2007. Did our ancestors speak a holistic protolanguage? *Lingua* 117: 579–604.

Teleki, G. 1975. Primate subsistence patterns: Collector-predators and gatherer-hunters. *Journal of Human Evolution* 4:125–184.

———. 1981. The omnivorous diet and eclectic feeding habits of chimpanzees in Gombe National Park, Tanzania. In R.S.O. Harding and G. Teleki, eds., *Omnivorous primates: Hunting and gathering in human evolution*, 303–343. New York: Columbia University Press.

Terrace, Herbert S. 1979. *Nim*. New York: Knopf.

Terrace, Herbert S., L. A. Pettito, R. J. Sanders, and T. G. Bever. 1979. Can an ape create a sentence? *Science* 200: 891–902.

Thomason, Sarah Grey. 2005. Typological and theoretical aspects of Hungarian in contact with other languages. In E. Fenyvesi, ed., *Hungarian language contact outside Hungary*, 11–29. Amsterdam: Benjamins.

Thorn, A. S., and S. E. Gathercole. 1999. Language-specific knowledge and short-term memory in bilingual and non-bilingual children. *Quarterly Journal of Experimental Psychology A* 52: 303–324.

Tipler, Frank J. 1980. Extraterrestrial intelligent beings do not exist. *Quarterly Journal of the Royal Astronomical Society* 21: 267–281.

Tomasello, Michael. 2000a. Do young children have adult syntactic competence? *Cognition* 74: 209–253.

———. 2000b. The item-based nature of children's early syntactic development. *Trends in Cognitive Sciences* 4: 156–163.

———. 2001. First steps toward a usage-based theory of language acquisition. *Cognitive Linguistics* 11: 61–82.

———. 2009. Universal grammar is dead. *Behavioral and Brain Sciences* 32: 470–471.

Trivers, R. L. 1971. The evolution of reciprocal altruism. *Quarterly Review of Biology* 46: 35–57.

Tse, L. 2001. Resisting and reversing language shift: Heritage-language resilience among U.S. native biliterates. *Harvard Educational Review* 7: 676–709.

Tulving, Endel. 1984. Precis of *Elements of episodic memory*. *Behavioral and Brain Sciences* 7: 223–238.

Uexküll, J. von. 1910. Die Umwelt. *Die neue Rundschau* 21: 638–649.

———. 1926. *Theoretical biology*. New York: Harcourt, Brace.

Ungar, Peter S., Frederick E. Grine, and Mark F. Teaford. 2006. Diet in early Homo: A review of the evidence and a new model of adaptive versatility. *Annual Review of Anthropology* 35: 209–228.

Uriagereka, Juan. 1999. Gloss on Chomsky: Derivation by phase. http://www.punksinscience.com/kleanthes/courses/UK03S/OSPS/NC JU DbP 1.pdf.

Valian, Virginia. 1981. Linguistic knowledge and language acquisition. *Cognition* 10: 323–329.

———. 1986. Syntactic categories in the speech of young children. *Developmental Psychology* 22: 562–579.

Vangsnes, Øystein Alexander. 2005. Microparameters for Norwegian wh-grammars. *Linguistic Variation Yearbook* 5: 187–226.

Venter, J. C., et al. 2001. The sequence of the human genome. *Science* 291: 1304–1351.

Wallace, Alfred Russel. 1869. Sir Charles Lyell on geological climates and the origin of species. *Quarterly Review* 126: 359–394.

Wassle, H. 2004. Parallel processing in the mammalian retina. *National Review of Neuroscience* 5: 747–757.

Watanabe, S. 1993. Object-picture equivalence in the pigeon: An analysis with natural concept and pseudoconcept discriminations. *Behavioural Processes* 30: 225–232.

Weibel, Ewald R. 1998. Symmorphosis and optimization of animal design: Introduction and questions. In E. R. Weibel, C. R. Taylor, and L. Bolis, eds., *Principles of animal design: The optimization and symmorphosis debate*, 1–10. Cambridge, U.K.: Cambridge University Press.

Weinreich, Uriel, William Labov, and Marvin Herzog. 1968. Empirical foundations for a theory of language change. In W. P. Lehmann and Y. Malkeil, eds., *Directions for historical linguistics: A symposium*, 95–188. Austin: University of Texas Press.

West-Eberhard, Mary Jane. 2003. *Developmental plasticity and evolution*. Oxford: Oxford University Press.

Wexler, Kenneth. 1998. Maturation and growth of grammar. In W. C. Ritchie and T. K. Bhatia, eds., *Handbook of child language acquisition,* 55–110. San Diego, Calif.: Academic Press.

White, Lydia. 1985. The "pro-drop" parameter in adult second language learning. *Language Learning* 35: 47–62.

Williams, G. C. 1966. *Adaptation and natural selection.* Princeton, N.J.: Princeton University Press.

———. 1992. Gaia, nature-worship, and biocentric fallacies. *Quarterly Review of Biology* 67: 479–486.

Wilson, Edward O. 1962. Chemical communication in the fire ant *Solenopsis Saevissima. Animal Behavior* 10: 134–164.

———. 1972. Animal communication. In W.S.-Y. Wang, ed., *The emergence of language: Development and evolution,* 3–15. New York: Freeman.

Wilson, R. A., and F. C. Keil, eds. 2001. *The MIT encyclopedia of the cognitive sciences.* Cambridge, Mass.: MIT Press.

Winford, Donald, and Bettina Migge. 2007. Substrate influence on the emergence of the TMA systems of the Surinamese creoles. *Journal of Pidgin and Creole Languages* 22: 73–99.

Wrangham, R. W. 1995. Ape culture and missing links. *Symbols,* Spring: 2–9, 20.

Wray, Alison. 1998. Protolanguage as a holistic system for social interaction. *Language and Communication* 18: 47–67.

———. 2000. Holistic utterances in protolanguage: The link from primates to humans. In C. Knight, M. Studdert-Kennedy, and J. R. Hurford, eds., *The evolutionary emergence of language,* 285–302. Cambridge, U.K.: Cambridge University Press.

Wylie, Jonathan. 1995. The origins of Lesser Antillean French Creole: Some literary and lexical evidence. *Journal of Pidgin and Creole Languages* 10: 77–126.

Yang, C. D. 2002. *Knowledge and learning in natural language.* Oxford: Oxford University Press.

Yip, Virginia, and Stephen Matthews. 2000. Syntactic transfer in a Cantonese-English bilingual child. *Bilingualism: Language and Cognition* 3:193–208

Young, Richard W. 2003. Evolution of the human hand: The role of throwing and clubbing. *Journal of Anatomy* 202: 165–174.

Yule, George. 2010. *The study of language.* 4th edition. Cambridge, U.K.: Cambridge University Press.

Zahavi, A. 1975. Mate selection—A selection for a handicap. *Journal of Theoretical Biology* 53: 205–214.

———. 1977. The cost of honesty (further remarks on the handicap principle). *Journal of Theoretical Biology* 67: 603–605.

Zentall, Thomas R., Edward A. Wasserman, Olga F. Lazareva, Roger K. R. Thompson, and Mary Jo Rattermann. 2008. Concept learning in animals. *Comparative Cognition and Behavior Reviews* 3: 13–45.

Zhou, Z. 2004. The origin and early evolution of birds: Discoveries, disputes, and perspectives from fossil evidence. *Naturwissenschaften* 91: 455–471.

Zipf, George K. 1935. *The psychobiology of language.* New York: Houghton-Mifflin.

Zurowski, Wirgiliusz. 1992. Building activity of beavers. *Acta Theriologica* 37: 403–441.

Acknowledgments

The longer one lives, the more apparent it becomes that no one ever does anything alone, and more particularly that the larger the scope of the endeavor, the greater the extent to which the work of others contributes to everything one accomplishes.

First in any list of acknowledgments must come the person to whom this book is dedicated, my wife, Yvonne. Without her constant presence in my life and her loving care, it is highly unlikely that I would have lived long enough to complete these pages.

Second must come my colleague, friend, and collaborator over the past two decades, Eors Szathmary. Most of what I know about biology is due to Eors, although he cannot be held responsible for any of my misunderstandings or any misuses I might have made of that knowledge.

Third is Noam Chomsky. This might seem surprising, since most of my current thinking runs dead counter to most of his current thinking. However, if it were not for his work, I could never have commenced the long and tortuous journey of which this book hopefully represents the final (or at least, semifinal) stages.

I would like to thank Tor Afarli and Britt Maehlum for inviting me to the Conference on Language Contact and Change held at Trondheim, Norway, on September 22–25, 2010. This conference served as a catalyst; it forced me to return to issues I had not dealt with since the beginning of my career and that I had thought irrelevant to current concerns. Gradually, half-formed ideas came into focus, cohered, and fell into place. The result was this book.

Several people read draft chapters, in particular Norbert Hornstein, Michael Studdert-Kennedy, and John Odling-Smee. I am deeply grateful for their incisive and insightful comments and criticisms, some of which I incorporated, others I probably should have.

Other people, not all of whose names I can even remember, have influenced my work in a variety of ways. Among them are Enoch Aboh, Michael Arbib, Robert Blust, Hagit Borer, Rudy Botha, Andrew Carstairs-McCarthy, Paola Cepeda, Samuel Epstein, Tecumseh Fitch, Tom Givon, Ian Hancock, Jim Hurford, Kevin Laland, Gary Marcus, and Sarah Roberts. My apologies to the many I am sure I have omitted.

At Harvard University Press, I was fortunate in having as my editor Michael Fisher, whose laid-back style was perfect for me. His acute criticism of early drafts saved me from going off in a number of wrong directions. Without his guidance, the book would never have assumed the shape it eventually did. I would also like to thank Heidi Allgair, Lauren Esdaile, and all the others at the Press for their expeditious and supremely efficient handling of the publishing process.

Index

acquisition, language, 9, 14, 23, 32, 185–187, 235, 250; and birdsong, 46–47, 49; of creole languages, 213–215; of empty categories, 144–145; one-word stage, 190, 194–196, 198; and parameters, 26–27, 29–30, 183; second-language, 123, 222, 224, 228, 230, 235–236, 250; two-word stage, 196–199, 206, 210
agreement, 149–150, 163, 167–168, 172
analysis: derivational, 34–36, 38; representational, 34–36
animal communication, 10, 11, 57, 74–75, 87, 92; and language precursors, 59–60
ants, 11, 66, 82–83, 85, 87, 94–96, 260
apes, 1–3, 56–57, 62, 70, 258, 272; culture in, 66; general intelligence of, 7; "language"-trained, 77, 120, 124–125, 189; and problem solving, 76–77; and protolanguage, 105, 122–123, 124, 190
arcuate fasciculus, 117
automaticity (of language), 162, 175, 261

Baldwin effect, 114, 115
Bantu languages, 13, 133
Basque, 171
bat, ghost, 107
bats, 6, 44, 46, 53, 63–64, 79; echolocation in, 43, 45, 48–49, 102

beavers, 6, 45, 51, 102; and dams, 53, 65, 67, 194
bees, 11, 82–83, 85, 87, 94–96, 260
behaviorism, 42
bilingualism, 228–229, 234, 253–255; native, 235
binding theory, 27–30, 135, 138–139, 192
biolinguistics, 81, 137
bioprogram, 219, 224, 238
birdsong, 46–49
Boeckx, Cedric, 27, 37, 39, 162, 225, 270
bottleneck, developmental, 191, 194, 202
brain, 2, 11–13, 23, 28, 73, 82, 91, 113, 129,148; of children, 208; dorsal and ventral pathways in, 125; economy in, 98, 101, 118, 153, 157, 162; and grammar, 21–22, 33–35, 38, 40, 106, 108, 115–117, 260; and hierarchical structure, 162–163; and memory, 161; neutrality of, 150, 262; as parallel processor, 36; plasticity of, 117; (pre)human, 45, 97, 105, 130, 189, 260; random activity in, 98–99, 266–267; scanning, 125–126; self-organization of, 12, 106–108, 115–119, 260; and sentence construction, 120–122, 130–134, 140–150; size, 82, 93, 96, 100–101; and vision, 124; and words, 100, 123, 130

319

case, 13, 150, 159, 172, 182; in acquisition, 190; as boundary marker, 173; case systems, 169–170. *See also* ergativity
categories, 74, 77, 79–80, 93–94; empty, 25, 29, 144–147; functional, 149, 180–181, 213; linguistic, 19, 148, 157–160, 171, 181, 208; role-based, 78, 264
causatives. *See* verbs: causative
c-command, 25, 134–140
Chinese, 13, 158, 166, 186, 200, 244; of Hawaiian children, 254
Chinese Pidgin English, 249
Chomsky, Noam, 6, 16, 20–21, 22, 24, 25–26, 27–36, 54–55, 62, 74, 112, 116, 152; and the bioprogram, 224–225; and language acquisition, 9, 185–186; and the Minimalist Program, 28–31, 131–132, 137, 180
citation dates, 234–223
cognition, 42–45, 74–80, 82, 88, 97–99, 259, 268–270; and language, 38–39, 263–265
cognitive gap, 5, 259, 263, 272
cognitive science, 31, 39, 45, 48
comparative method, 48–49, 69–70
competence-performance distinction, 21–22, 140
concepts, 12, 88, 92–94, 119, 162, 264, 270; linkage between, 80, 124, 130, 150, 261, 267–268; in nonhumans, 74–80; structure of, 96–104
consciousness, 3, 44, 208, 259; stream of, 266–268
constituent structure, 110, 112
continuity, evolutionary, 5, 42, 62, 64
continuity hypothesis (language), 192–193, 216
convergence, evolutionary, 69, 81–82, 86, 272
convergence zone, 97
cooperation, 82, 85–89
covert movement, 136–140, 207, 209, 212
creationism, 4, 6, 55, 65, 275
creole continuum, 220–221
creole languages, 123, 146, 159, 166, 178, 218; and acquisition, 201, 207, 210, 213–216; continuum of, 220–222; criticisms, 224–242; in Hawaii, 249–255; plantation cycle, 222–224; Sranan and Saramaccan, 242–249; universalist case, 255–257
culture, 11, 44, 66–67, 152, 268
cut-marks, 84

Darwin, Charles, 1, 3–4, 6, 8, 14, 67, 76, 90; and "continued use," 262–263, 265; and musical protolanguage, 55; neo-Darwinism, 64; neural Darwinism, 131–132
Deacon, Terrence, 10, 96, 152, 157
design features, 50, 82
digestion, 33, 44
displacement, 11, 85, 89, 104, 120, 260; and confrontational scavenging, 87; as "design feature," 82; and symbolism, 91–92, 94–96, 99; syntactic, 144
dreaming, 267
dualism, 102, 110, 180
Dunbar, Robin, 54, 63

echolocation, 6, 11, 43, 45, 48–49, 64, 102, 107
elephant, 11, 56, 88
empiricism, 31–32, 41, 144–145, 256; and language acquisition, 186, 205, 206, 212, 214; and nativism, 10, 13–14, 46, 116, 151, 257, 274–275
English, 163, 167, 186, 250; acquisition of, 200, 202–212; related creoles, 240–241
episodic memory, 128–129, 131, 160
ergativity, 171
eusociality, 82, 85
evidentiality, 160, 201

evo-devo (evolutionary-developmental biology), 51–53, 58, 64; and niche construction, 65
evolutionary convergence. *See* convergence, evolutionary

finality, 137–140, 145, 147–148
Finnish, 13, 170
Fitch, Tecumseh, 47, 54–55, 60–61
flight, origin of, 71
foraging, 11, 83, 85; optimal theory of, 83; unit size in, 87
French, 92, 133, 162; acquisition of, 206–209; related creoles, 178, 182, 233, 236
functional reference. *See* reference

game theory, 86
gender, 166–168, 201
generative grammar, 16–37, 110–111, 134–135, 138–142, 180–181; and acquisition, 187, 206–207; and creole languages, 238, 256; and evolution, 115; Principles and Parameters, 24–28, 186, 225; Standard Theory of, 19–24, 26. *See also* minimalism
genome sequencing, 51
grammaticizable distinction, 201–202
grammaticization, 161, 165–166, 169, 172–173, 222

Haitian, 178, 226, 233, 235–236, 256; and population figures, 222, 229–230
Hauser, Marc, 7–8, 49–55, 63–64, 60, 128
Hawaii, 220, 222–224, 227–230, 232, 237, 242, 245
Hawaiian (language), 134, 223, 232–233, 240, 244
Hawaiian Creole English, 213, 218, 249–255
Hawaiian Pidgin English, 234, 240, 249–255
Hebb's Rule, 101, 123

Homo erectus, 60, 68, 84–85, 116
Homo habilis, 60, 68, 84–85
homology, 49; deep homology, 52–53, 55
Hornstein, Norbert, 31–32, 37, 134–135, 184
Hungarian, 13, 170, 191

icons, 92–95
indices, 92–95
instinct, 21, 44–45, 66–67, 82; "instinct to learn," 45–47
intelligence, 1, 6–8, 76, 93, 103, 262, 268; extraterrestrial, 272–274; social, 7, 62; specialized, 7–8
intelligent design, 4, 6, 55, 275
intraspecific variation, 43

Jiwarli, 111–112, 172
Joyce, James, 265–266; "Joyce factor," 263, 265

laborers, indentured, 249–250
language, polysynthetic, 168, 190–191, 202
language acquisition device. *See* acquisition, language
language of thought, 14, 103, 115, 153
language organ, 32, 38, 126
Larsonian shells, 136
learning, 9–10, 31, 39, 44, 66–67, 185–188, 206, 230; of creoles, 222; delays in, 193; general-purpose mechanisms, 112; inductive, 210–211; item-based, 205; of polysynthetic languages, 191, 200, 212; second-language, 123; statistical, 196; vocal, 57; of words, 191, 197, 199, 209, 211, 216
lexicon, 30, 123, 148, 155, 184; of pidgin, 229; of Sranan and Saramaccan, 244–247
lexifier, main, 218, 220–223, 227, 233, 236, 239, 241–242, 256; Portuguese as, 256; of Surinamese creoles, 248
life, origin of, 73

macaronicity, 236–237, 242, 244, 249, 251
maturation hypothesis, 192–193, 202, 216
meat, 84, 86, 88; increased consumption, 89; sharing of, 85
megafauna, 83, 88
memory, 77; autobiographical, 101; episodic, 125–129, 131, 160
merge, 29–31, 34–35, 40, 54,111, 180–181; or attach, 132–135, 147, 159; in evolution, 105–106, 124
micro-parameter. *See* parameters
mind, 1–4, 16, 23, 76, 94, 261–262, 276; mind/brain, 28, 32, 81, 146; "mind-independent objects," 74; in non-humans, 88, 189; theory of, 93
minimalism, 28–31, 33–35, 39, 111, 131, 135, 141, 154, 180–181, 187, 225
Mohawk, 168
Morisyen, 178, 214, 231–232, 236
morphology, 30, 34, 136, 170, 191, 200–202; Turkish, 178
movement, 19–20, 26, 29–30, 122, 141–142, 144; copy theory of, 40, 111; covert, 136–140, 212; and "islands," 35; subject-raising, 208; verb-raising, 182
Mufwene, Salikoko, 189, 218–219, 222, 237

nativism, 31, 32, 186, 272; and acquisition, 186, 206, 212, 214; and empiricism, 10, 13–14, 46, 116, 151, 257, 274–275
natural selection, 1–2, 11, 53, 90, 265, 271, 275; and the brain, 107, 118, 153; and language, 11, 55, 62, 152
negation, 203, 206–208
neural linkage, 80, 101, 123, 261
niche construction, 53, 58, 64–69, 71, 96, 101; and language, 114–116; and scavenging, 83, 85, 90, 259, 264
nouns, 20, 63, 111, 126–127, 130–132, 148, 164, 180; acquisition of, 188, 194–195, 200; and agreement, 163; classifiers, 167; and gender, 166–167; and "islands," 142; referential, 139–140; storage of, 119; as verbs, 215; and word order, 27, 156–157, 197

parameters, 26–27, 29, 182–184; micro-, 183; unmarked settings of, 225–226
Penn, Derek, 5, 77–78, 263–264, 269
phonology, 50, 130, 132, 155–156, 161, 165; acquisition of, 194, 195–196, 200
pidgin, 104–105, 122–123, 125, 189–191, 193, 220, 236–237; and compounding, 248–249; and dominance, 254; in Hawaii, 225–229, 234, 240, 251, 255; macaronicity in, 244, 249; on plantations, 222–224; in Suriname, 242–243, 245–246
pidgin-creole cycle, 218
Pidgin Hawaiian, 250
pigeons, 76
plantations, 220–224, 227, 238, 250; children in, 229–230; in Suriname, 242–243, 245
Plato, 20–21, 31, 37, 186–187
pressure, selective, 55, 62–63, 69, 89, 115–116, 262
primates, 56–57, 61–62, 68–70, 78, 83, 89, 273; cooperation among, 85–86; foraging by, 84; grasp of, 153; and precursors, 58–59; "primate-centric" approach, 56, 58; reciprocity in, 128–129
priority, 137–140, 142, 145, 147–148
protolanguage, 95, 97, 103–106, 121, 124, 127, 147, 268; cause of, 115–116, 162; and child language, 188–192; holophrastic, 106; musical, 55, 61; and pidgin, 222
punctuated equilibrium, 68

questions, 19, 40, 122, 145; acquisition of, 210–212; question-words, 141, 226, 239–240

ravens, 82, 85
reciprocal altruism, 86, 128, 131, 259
recruitment, 83, 85–86, 88, 95, 162
reference, 24, 28, 92–93, 139–140, 144–147; functional, 57, 79, 88; ostensive, 195. *See also* displacement
relationships, higher-order, 78, 264
Riau Indonesian, 161, 164
Roberts, Sarah, 223, 227, 229, 232–234, 236, 250–256
robots, 95

Saramaccan, 146, 215, 240, 242–249
savanna, 84, 87, 116, 276
scala naturae, 55, 60
scavenging, confrontational, 83–90, 96, 105, 115–116, 264; and cooperation, 259; and displacement, 99–100; and recruitment, 115, 162: and throwing, 130
scrub-jay, 76, 128
self-organization, 12, 106–107, 117–118, 260
sequence marker, 240
serial verb, 146, 204, 215, 226, 231–233
Seselwa, 146, 214–215, 231–232
Seth, 201, 216
Shakespeare, 2–3
slavery, 222–224, 231–232, 242–243; and indentured labor, 249–250
Spanish, 26, 30, 92, 167–168, 171–172, 174–175, 247
speciation, 66, 67–69, 114, 178
speech, 8, 3, 109,155–156; caregiver, 210–212; of infants, 193; speed of, 36, 122–123; and substitution errors, 132
spiders, 6, 43–45, 46, 48, 50, 53
Sranan, 226, 233, 234, 236, 240–249
subset principle, 204, 206
sub-song, 46
substrate language, 226–228, 230–235, 238, 251, 254
Suriname, 222, 230, 240, 242–243, 245, 247–248

Swahili, 166–167
switch-reference, 168,175
symbolism, 12, 92–95, 260, 264; and brain storage, 115, 116; and concatenation, 92, 120–121, 124, 149, 199; and concepts, 77; continued use of, 265; and culture, 66; and displacement, 91–92, 95, 99, 106; and protolanguage, 103, 105; and reference, 93–94
syntactic island, 141–144
syntactic spurt, 216
syntax, 6, 13, 16–36, 109, 124, 147, 212; in acquisition, 187, 191, 206, 219; bare, 111, 186; in brain, 126, 139; claimed for animals, 60; emergence of, 104–105; and maturation, 192–193; and nativist-empiricist debate, 274; and phonological change, 156; and predication, 125; in thought, 103, 129–130, 153, 162; universal, 134. *See also* X-bar theory

telegraphic speech, 46, 200–203, 207
tense, 155, 156, 163, 167; in acquisition, 188, 201, 207–208
tense-mood-aspect (TMA) system, 158–161, 201, 208; in creoles, 214, 226, 229, 233, 235, 239–241
thinking, 12–13, 101, 116, 150, 260, 265; human versus nonhuman, 124; off-line, 79–80, 101–103; unconscious, 268
throwing, 12, 88, 129–130, 153, 260–261
Tomasello, Michael, 109, 186, 205
tools, 50, 89, 96, 259
Turkish, 190–191, 202–203, 212

Umwelt, 58, 94
underspecification, 39, 149, 156, 177, 241; in acquisition, 203; in creolization, 231, 238–239; repair of, 158, 164, 176, 179, 256

uniqueness, human, 49–50, 59, 63, 89
Universal Grammar (UG), 20, 22, 39, 109,134, 203, 224–225
universal language, 13, 105, 110–112; and change, 157; and creoles, 220, 238, 241–242; human, 62, 131; "impossibility" of, 113–114; of TMA, 158, 160–161

verbs, 20, 26, 63, 137, 166; in acquisition, 188, 195–196, 206–215; algorithms for, 148; causative, 205–206; and complementizers, 175–176; in creoles, 178; as heads, 126–129, 130; Latin, 156; as nouns, 164; and phases, 137; raising, 182; storage of, in brain, 119; and thematic roles, 131, 133, 169–172; and TMA, 158–159; in VP shells, 136; and word order, 157–158, 163, 197. *See also* serial verb
vervet monkeys, 57, 60, 74–75, 92–93, 120, 152

Wallace, Alfred Russel, 1–3, 12, 272; Wallace's problem, 4–6, 14, 87–91, 259, 262, 271
word order, 125, 156–157, 161, 197; in acquisition, 210; in creoles, 229–238; "free," 111, 172; and raising, 209; SVO, 163, 169; and topic-comment, 190
words, 6, 12–13, 34–35, 45–46, 99–102, 198, 268; in acquisition, 187–189, 194–199, 202, 209; classes, 27, 119, 132, 181; and concepts, 102; and constituency, 112; in creoles, 244–248; merging of, 35, 121–124, 133; and morphemes, 34; onomatopoeic, 92; in pidgins, 236; precursors of, 60; in protolanguage, 104, 106; reference of, 92–93, 127. *See also* brain
words, question. *See* questions

X-bar theory, 29, 181

Zipf's Law, 198
Zuni, 168